Salivary Glands

Frontiers of Oral Biology

Vol. 14

Series Editor

Paul Sharpe London

Salivary Glands

Development, Adaptations and Disease

Volume Editors

A.S. Tucker London
I. Miletich London

34 figures, and 12 tables, 2010

Basel · Freiburg · Paris · London · New York · Bangalore · Bangkok · Shanghai · Singapore · Tokyo · Sydney

Frontiers of Oral Biology

Abigail S. Tucker, DPhil
Isabelle Miletich, DDS, PhD
King's College London
Craniofacial Development Department
Floor 27, Guy's Tower
Guy's Hospital
London SE1 9RT (UK)

Library of Congress Cataloging-in-Publication Data

Salivary glands : development, adaptations, and disease / volume editors, A.S. Tucker, I. Miletich.
p. ; cm. -- (Frontiers of oral biology, ISSN 1420-2433 ; v. 14)
Includes bibliographical references and indexes.
ISBN 978-3-8055-9406-6 (hard cover : alk. paper)
1. Salivary glands. 2. Salivary glands--Diseases. I. Tucker, A. S. (Abigail S.) II. Miletich, I. (Isabelle) III. Series: Frontiers of oral biology, v. 14. 1420-2433 ;
[DNLM: 1. Salivary Glands. 2. Salivary Gland Diseases. W1 FR946GP v.14 2010 / WI 230 S1667 2010]
RC815.5S24 2010
616.3′16--dc22

2010003731

Bibliographic Indices. This publication is listed in bibliographic services, including Current Contents® and Index Medicus.

www.karger.com
Printed in Switzerland on acid-free and non-aging paper (ISO 9706) by Reinhardt Druck, Basel
ISSN 1420–2433
ISBN 978–3–8055–9406–6
e-ISBN 978–3–8055–9407–3

Contents

VII **Preface**
Tucker, A.S. (London)

1 **Introduction to Salivary Glands: Structure, Function and Embryonic Development**
Miletich, I. (London)

21 **Salivary Gland Adaptations: Modification of the Glands for Novel Uses**
Tucker, A.S. (London)

32 **Genetic Regulation of Salivary Gland Development in *Drosophila melanogaster***
Pirraglia, C.; Myat, M.M. (New York, N.Y.)

48 **Extracellular Matrix and Growth Factors in Salivary Gland Development**
Sequeira, S.J. ; Larsen, M.; DeVine, T. (Albany, N.Y.)

78 **Lumen Formation in Salivary Gland Development**
Wells, K.L.; Patel, N. (London)

90 **Epithelial Stem/Progenitor Cells in the Embryonic Mouse Submandibular Gland**
Lombaert, I.M.A.; Hoffman, M.P. (Bethesda, Md.)

107 **Salivary Gland Regeneration**
Carpenter, G.H.; Cotroneo, E. (London)

129 **Salivary Gland Disease**
Thomas, B.L.; Brown, J.E.; McGurk, M. (London)

147 **Author Index**

148 **Subject Index**

Preface

Salivary glands are vital parts of the oral cavity, defects in which can cause major disruptions to our lifestyles. This book brings together basic science researchers and clinicians to produce a review of the latest developments in salivary gland research. The book is divided into four broad areas, which chart our current understanding of salivary gland morphology, development, regeneration and disorders.

In the chapters by Miletich and Tucker, the salivary glands are introduced and unusual adaptations are investigated. These chapters aim to introduce the complex nature of salivary glands and highlight their huge variation in size, shape and function across the animal kingdom. In the following three chapters by Pirraglia and Myat, Sequeira et al., and Wells and Patel, the development of the salivary gland is addressed from specification to branching morphogenesis and lumen formation. Here data is brought together from two diverse animal models, *Drosophila* and mouse, to provide an understanding of the basic steps of salivary gland development. These chapters introduce the genes and complex signalling pathways that direct development as the gland grows from initiation to differentiation. The book then turns to the prospect of regenerating salivary glands in adult tissue in the contributions by Lombaert and Hoffman and Carpenter and Cotroneo. Lombaert and Hoffman focus on the location of stem cells in embryonic glands, providing exciting new data on the role of growth factors in determining cell fate. The article by Carpenter and Cotroneo moves to a rat model of gland regeneration to study the molecular triggers and morphological changes involved in regeneration. These chapters highlight new areas of research that may shape the way salivary gland disorders are treated in the future. Finally, Thomas et al. look at the disorders of salivary glands from a clinical perspective, detailing how salivary gland disorders come about, and what techniques are being developed to help treat patients. It is hoped that together these chapters will provide an intriguing overview of salivary gland development, disorders and treatment, which will be of interest to developmental biologists, anatomists and clinicians.

Abigail S. Tucker

Tucker AS, Miletich I (eds): Salivary Glands. Development, Adaptations and Disease.
Front Oral Biol. Basel, Karger, 2010, vol 14, pp 1–20

Introduction to Salivary Glands: Structure, Function and Embryonic Development

Isabelle Miletich

Department of Craniofacial Development and Orthodontics, Guy's Hospital, London, UK

Abstract

Salivary glands are a group of organs secreting a watery substance that is of utmost importance for several physiological functions ranging from the protection of teeth and surrounding soft tissues to the lubrication of the oral cavity, which is crucial for speech and perception of food taste. Salivary glands are complex networks of hollow tubes and secretory units that are found in specific locations of the mouth and which, although architecturally similar, exhibit individual specificities according to their location. This chapter focuses on the embryonic development of vertebrate salivary glands, which has been classically studied in the mouse model system since the 1950s. We describe here where, when and how major salivary glands develop in the lower jaw of the mouse embryo. Key mechanisms involved in this process are discussed, including reciprocal tissue interactions between epithelial and mesenchymal cells, epithelial branching morphogenesis and coordinated cell death and cell proliferation.

Salivary Glands as Multifunctional Organs

Terrestrial animals possess salivary glands, which are exocrine glands producing saliva, a watery substance that is excreted in the mouth. Salivary glands are either absent or very rudimentary in animals living in water [1]. For example, these glands are absent in aquatic animals such as fish, whose oral cavity fills with considerable amounts of liquid upon opening of the jaws. However, one pair of salivary glands exists in lampreys [2], parasitic jawless vertebrates that feed by boring into the flesh of various species of bony fishes to suck their blood. Lampreys have a sucking mouth that does not let water get into the oral cavity.

Saliva performs a wide array of physiologic and protective functions, some related to its fluid properties and others to its specific content of a variety of molecules [3]. Being a liquid, saliva primarily lubricates the oral mucosa lining the inside of the mouth and moistens food bites. As such, it cleans the oral cavity by flushing away food debris and bacteria, helps with mastication and swallowing of the food bolus, facilitates speech, and, last but not least, allows taste perception by solubilizing food chemicals, an essential step for the stimulation of receptor cells of the taste buds. Although mostly composed of water, saliva also contains electrolytes and an incredible variety of proteins and peptides fulfilling numerous functions. Specific components actively secreted in the saliva are key to maintaining the good health of the oral cavity. Saliva protects the teeth through the presence of negatively charged proteins that bind to hydroxyapatite minerals on the enamel surface of tooth crowns. Through its high bicarbonate concentration, it buffers acids produced by the dental plaque bacteria when carbohydrates are fermentated, thereby preventing tooth decay. Saliva also provides protection to the oral mucosa lining the inside of the mouth, via an array of antimicrobial agents including secretory immunoglobulin A, lysozyme and lactoperoxidase. In addition to its defensive role, saliva also initiates the digestion of starches and a small fraction of triglyceride lipids through α-amylase and lipase enzymes respectively. However, these two enzymes are considered to be of minor significance in healthy individuals since they are rapidly inactivated by gastric acidity. Apart from components having an obvious function, saliva also exhibits a tremendous variety of biologically active proteins in the form of growth factors and other small peptides, whose function remains largely unknown [4, 5]. The paramount importance of saliva is illustrated by the plethora of problems experienced by individuals with non-functional salivary glands, which produce decreased volumes of saliva leading to dry mouth (xerostomia). These include oral infections, dental caries, mastication and swallowing problems, loss of taste, pain or discomfort on eating or talking that have detrimental effects on the quality of life. Salivary gland disorders and their treatments are described by Thomas et al. [pp. 129–146] in this book.

Although salivary glands are thought of as organs whose function is mostly related to the maintenance of the oral cavity and the digestion of food substances, their secretions, in various animal species, have evolved to perform functions other than those previously cited. In particular, some birds, insect larvae, reptiles and small mammals have developed specialized uses for saliva that are described in detail by Tucker [pp. 21–31].

Structures and Cell Types of Adult Salivary Glands

Mature adult salivary glands consist of a parenchyma, the glandular secretory tissue, and a stroma, which is the supporting connective tissue. The parenchyma is composed of secretory units called secretory 'endpieces', which are connected to the oral cavity through a network of ducts. Salivary endpieces consist of secretory cells organized in round clusters, termed acini, or tubular clusters, called tubules. Secretory endpieces belong to three different types, mucous, serous, or seromucous, depending on the composition of their secretions. Mucous secretions are rich in complex carbohydrates, found as chains attached to mucin proteins, which represent most of the mass of these glycoproteins. Serous secretions are rich in proteins, with a notable absence of mucin proteins, while seromucous secretions are a combination of both serous and mucous secretions. Consistency of acinar cell secretions varies with their composition; serous secretions are watery, whereas mucous secretions are viscous and adhere to oral structures, accounting for most of the lubrication effect of saliva. Interestingly, the different cell types found in secretory endpieces can be identified by their characteristic tissue organization and specific cell structure that are easily distinguished by histological stainings. Serous cells are pyramidal and as such form spheroidal clusters or acini. They display a large round central nucleus and small discrete apical granules that are darkly stained with haematoxylin. Mucous cells are columnar and organized in elongated tubular clusters. Resting mucous cells contain numerous large and close-packed granules that occupy their apical two thirds, pushing and flattening the nucleus at the base of the cell; their apical cytoplasm appears poorly stained with haematoxylin-eosin stain. In contrast, mucous cells are very distinctively stained blue with alcian blue staining. Secretory endpieces empty their secretions consecutively in intercalated ducts, striated ducts, excretory ducts and finally the main excretory duct that opens in the oral cavity. Intercalated ducts are lined by small cuboidal cells, striated ducts by columnar cells arranged in a simple or pseudostratified organization and excretory ducts are lined by a stratified columnar epithelium. The duct system is impermeable to water. However, it actively modifies the ionic content of the saliva in specific ductal areas, such as the striated ducts. Striated ducts are so called because the cells lining these ducts display basal striations due to cytoplasmic infoldings in which are located vertically aligned mitochondria that provide the energy necessary for active ion exchange at the apical membrane of these duct cells. Saliva production is a two-stage process. An isotonic plasma-like secretion is initially produced by secretory endpieces. When passing down the striated ducts, this primary fluid is rendered hypotonic as excess sodium ions are reabsorbed and potassium and bicarbonate ions are added. Although the cells located in secretory endpieces are the main producers of salivary proteins and glycoproteins, duct cells also secrete proteins. For example, striated ducts secrete immunoglobulin A and lysozyme. Many species of rodents, including mice and rats, exhibit an additional type of duct located between the intercalated and striated ducts and known as granular ducts or granular convoluted ducts [6]. The granules present in these cells are strongly stained by haematoxylin and basic dyes such as toluidine blue; they contain various growth factors and non-specific proteases. In humans, in which granular ducts are not present, these proteins are produced and secreted by striated ducts. Associated with secretory endpieces and proximal ducts (including intercalated ducts and, in rodents, granular ducts) are found myoepithelial cells. They are contractile cells, resembling smooth muscle cells, which are located between the basal lamina and the cytoplasmic membrane of either the cells of the secretory endpieces or the cells lining the salivary ducts. They interact with endpiece or ductal cells via desmosomes. Their shape depends on their location; at the level of endpieces, they appear as stellate, dendritic cells forming a basket around each endpiece, whereas in the wall of the ducts they are fusiform, with few cytoplasmic processes, and run parallel to the ducts. Contraction of myoepithelial cells helps secretory cells to discharge the content of their secretion granules, reduces the luminal volume of endpieces and ducts, resulting in increased salivary flow, and also aids to support and stabilize the glandular tissue against strong secretory pressures applied during periods of high saliva production. Other components of adult salivary glands include nerves, blood vessels and the fibrous connective tissue capsule covering each gland. The capsule projects septa into the salivary gland, dividing the parenchyma into lobules. Intercalated and striated ducts are located inside lobules, whereas excretory ducts are found in between lobules, within the connective tissue septa.

Different Salivary Glands, in Different Locations

Salivary glands are classically divided into major and minor salivary glands. Major salivary glands are large glands located at a distance from the oral mucosa, which empty their secretions in the oral cavity through long extraglandular ducts, whereas minor salivary glands are small secretory units contained inside the oral mucosa that open either directly in the mouth or indirectly through many short ducts. Unlike major salivary glands, minor salivary glands are not encapsulated by connective tissue, they are only surrounded by it. In humans, major salivary glands comprise three pairs of glands, namely the parotid, submandibular and sublingual glands, from which is derived 90% of the total saliva (fig. 1). In humans, parotid glands, which are the largest salivary glands, are located at the back of the mouth at the front and below of the ears. Saliva is excreted in the oral cavity through a 5 cm long duct (Stensen's duct) that opens opposite the second upper molar crown. Submandibular glands are located in the floor of the mouth in between muscle layers and touch the mandibular bone. They discharge their secretions each through a duct (Wharton's duct) opening in the floor of the mouth on the sublingual papilla located at the base of the lingual frenum posterior to the lower incisors. Submandibular secretions account for 70% of the saliva produced by major salivary glands. Sublingual glands are the smallest of the major salivary glands; they lie in the floor of the mouth just beneath the oral mucosa closer to the midline than the submandibular glands and empty through 8–20 excretory ducts that open under the tongue on the sublingual fold. Sometimes anterior acini of the sublingual gland drain into a main excretory duct (Bartholin's duct) that terminates with or near the orifice of the submandibular duct. In addition to these three pairs of large salivary glands, 600–1,000 minor salivary glands are scattered in the oral mucosa of the hard palate, tongue, floor of the mouth, inner side of the cheeks and lips and oropharynx. Minor salivary glands are not found within the gingiva, anterior part of the hard palate and dorsal surface of the anterior part of the tongue. These glands contribute only a small portion (10%) of total salivary secretions. Although closely related, major salivary glands exhibit differences in their architecture and secretions. The parotid glands are flat glands spread over a large area

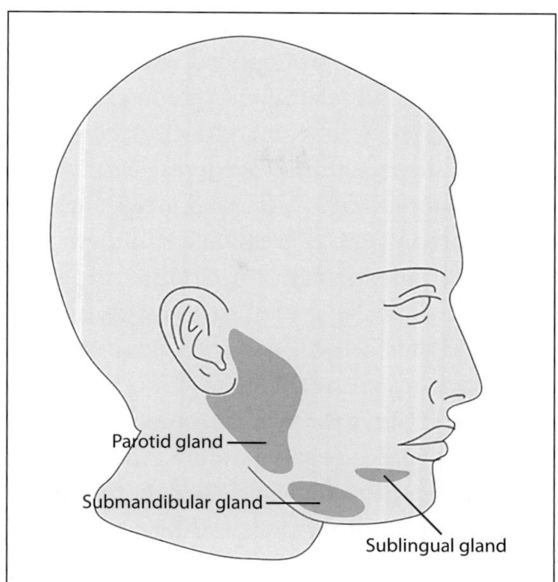

Fig. 1. Location of major salivary glands in humans.

with clearly visible epithelial branches, since these are separated from each other by large areas of mesenchymal tissue, whereas the submandibular and sublingual glands are globular, compact glands, with so densely packed epithelial branches that distal secretory endpieces touch each other, giving a 'bunch of grapes' appearance to these glands. Interestingly, each type of salivary gland produces a secretion with a distinctive composition suggesting that these organs function in a coordinated manner to generate a fluid adapted to environmental conditions. In humans, the parotid glands are exclusively serous, the sublingual glands mainly mucous in nature, and the submandibular glands seromucous mixed glands with a predominance of serous acini. In mice, the parotid, submandibular and sublingual glands are of the same type as in humans. In humans and mice, minor salivary glands are predominantly mucous in character, with exception of von Ebner's glands, a small group of salivary glands located on the posterior region of the tongue, beneath the circumvallate papillae, which are exclusively serous.

The largest volumes of saliva are by far produced by terrestrial mammals, which also display the largest salivary glands. The same major salivary gland configuration as described in humans can be observed in other mammals. An additional salivary gland, the zygomatic gland, which is located under the eye, has been described in carnivores including the domestic cat and dog, lagomorphs and some ruminants [7]. Rather than a proper salivary gland, it appears to be several dorsal buccal accessory glands grouped in a compact mass. This group of glands drains in the oral cavity through five major excretory ducts opening at the level of the last upper molar.

Variations in size, structure and secretions of salivary glands can be related to sex. In mice there is a pronounced sexual dimorphism of the submandibular gland. In males, submandibular glands are more voluminous, granular convoluted ducts containing secretory cells are more prominent and in mature glands, production of some enzymes such as aminotransferase can be as much as 10 times less than in females [8–10]. More recently, a microarray analysis has identified numerous sex-related differences in gene expression in all three major salivary glands of male and female mice [11]. Interestingly, importance and size of the different major salivary glands also appears to be related to the type of diet of mammalian species. For example, parotid glands, which, as serous glands, produce a fluid saliva that is mainly used to moisten food, are very developed in herbivores and less voluminous in carnivores. Horses and other mammals feeding on a diet of dried substances, such as hay, have indeed very large parotid glands and small submandibular glands. In contrast, other mammals such as anteaters, which are insectivores that use a long filiform tongue to catch ants and termites, exhibit unusually large submandibular glands and very small parotid glands. These animals collect insects by gluing them onto their tongue, which is coated with a very sticky saliva. This viscous saliva is secreted by the submandibular glands; it covers oral cavity structures including the tongue, and is therefore crucial for feeding. The voluminous submandibular glands extend along the long necks of these animals and adjacent chest area, where they even penetrate in the space between mammary glands, instead of being located in the floor of the mouth, as in humans. The type of diet also appears to influence the type of saliva secreted by the mixed submandibular glands; it is predominantly mucous in carnivores and mostly serous in herbivores and rodents.

Control of Saliva Production

Saliva production is continuous although the flow rate varies greatly during a day, following a circadian rhythm with an afternoon peak and a secretion near to zero during sleep [12]. Regulation of the volume and quality of saliva is achieved through the regulation of the activity of salivary effector cells, comprising myoepithelial cells, endpiece secretory cells and duct secretory cells, as well as through the regulation of the diameter of salivary gland blood vessels. Variations in salivary secretions are under neural control and more specifically under control of the autonomic nervous system (ANS) or vegetative nervous system, which is mainly concerned with the regulation of visceral functions and interactions with the internal environment. The ANS has two divisions, the parasympathetic autonomous system (PANS), aiming at homeostasis and conservation of body functions and the sympathetic autonomous system (SANS), providing answers to stress. PANS and SANS are both involved in the control of saliva production. Importantly they act synergistically, since stimulation of both PANS and SANS leads to an increase in saliva production. However, stimulation of these two nervous systems differs in their effects on the fluid volume and protein content of the saliva secreted. Parasympathetic stimulation, which is most active during the day whilst eating, leads to the production of large volumes of saliva with low protein content. These watery secretions are predominantly produced by serous endpieces of the parotid and submandibular glands under chemical stimulation of the acetylcholine neuromediator. Stimulation of sympathetic nerves, leading to the release of noradrenaline, results in the secretion of low volumes of a protein-rich saliva. Depending on the noradrenergic receptor subtype, viscous secretion (α-receptor type) or amylase secretion (β-receptor type) is released. This thicker, mucous-rich saliva is mainly produced by the sublingual gland and partly by the submandibular gland. Such a secretion happens in certain situations when fear, stress or anger are aroused, or during hard physical exercise. PANS and SANS also affect salivary gland secretions indirectly by innervating the blood vessels that supply the glands. Water, the major component of salivary secretions, is obtained from the lymph filling the lymph spaces adjacent to the secretory endpieces. Since lymph exudes from blood vessels, any effect on the permeability of local blood vessels has repercussions on saliva production. As a matter of fact, abundant saliva secretion is usually associated with an abundant blood supply. PANS and SANS exert an antagonistic action on blood vessel diameter, and therefore permeability; stimulation of PANS results in vasodilatation while stimulation of SANS causes vasoconstriction. Accordingly, the volume of saliva secreted is large following PANS stimulation and low after SANS stimulation. Parasympathetic innervation of the salivary glands is achieved through cranial nerves originating in the brainstem, which is the lower part of the brain contiguous with the spinal cord. Parasympathetic innervation of the parotid gland is achieved via the glossopharyngeal nerve (cranial nerve IX), following the tympanic branch of this nerve, which connects in the otic ganglion with second-order neurons that synapse with parotid gland cells. The sublingual and submandibular glands receive their parasympathetic input from the facial nerve (cranial nerve VII) through the chorda tympani branch of this nerve, which connects in the submandibular ganglion with second-order neurons that innervate these two salivary glands [13]. In addition to the control exerted by these two kinds of autonomic nerves, unconditioned reflex pathways triggered by sensory stimuli also stimulate saliva production. Olfactory and gustatory stimuli, mastication, are peripheral inputs capable of stimulating salivation.

The Mouse Submandibular Gland as a Model for Vertebrate Salivary Gland Development

The mouse submandibular gland is classically used as a model system to study vertebrate salivary gland development. It is a mixed gland, and as such, terminal differentiation of secretory cells of endpieces is related to cytodifferentiation of both the parotid and sublingual glands. In terms of early embryonic development, the submandibular, sublingual and parotid glands appear to follow similar morphogenetic events although these three glands start developing at different time-points. In mice, the submandibular gland (SMG) develops first, followed one day later by the sublingual gland (SLG). The parotid gland (PG) is last to develop and the minor salivary glands initiate their development even later. In humans, the SMG starts developing around week 6 of fetal life, the PG between the 6th and 7th week, followed by the SLG around the 8th week [14]. In mice, contrary to humans, the SMG and SLG develop in close association, ending up in the same anatomical location in adults where they share the same capsule of connective tissue. Despite the closeness of their developmental relationship, these two glands display in their mature state the same functional differences as those observed in humans; the adult mouse submandibular gland being predominantly serous and the adult mouse sublingual gland predominantly mucous.

Transgenic mice are very useful tools to understand the role of signalling pathways and molecules involved in the development of mammalian salivary glands (see table 1 for a list of salivary gland defects that have been observed in mouse mutants). Apart from mouse mutants, one important advantage of the mouse model is that the entire salivary gland rudiments of the SMG and SLG can be cultured in vitro in serum-free medium. The first organotypic cultures of mouse salivary glands were performed in the 1950s [15] and have since proved to be a very good model to study salivary gland development. In vitro development of these two glands recapitulates normal development in live embryos with few minor differences. Similar to other organ cultures, salivary gland development is slightly slower in vitro; when explants are cultured for more than 48 h, secretory endpieces appear to be grouped near the periphery of the gland whereas they are evenly scattered in vivo and finally, ducts undergo an abnormal dilatation in in vitro cultures that remains unexplained so far and hinders the study of regulation of lumen size in this culture system [15]. Despite these differences, advantages of this culture system are numerous. Importantly, salivary gland development can be observed. Epithelial branching can be readily followed in living explants and clear images of epithelial and mesenchymal tissues can be obtained in fixed preparations of these explants. Recent advances in cell-marking techniques have provided tools for live imaging of salivary gland development. Recombination of wild-type salivary mesenchyme with salivary epithelium taken from transgenic mice expressing Green Fluorescent Protein (GFP) [16] allows to follow epithelial morphogenesis and more specifically branching morphogenesis. Movements of individual epithelial cells have also been studied during branching morphogenesis by confocal time-lapse microscopy after labelling epithelial cells by microinjection of a GFP adenovirus construct [17]. Finally, in vitro culture of salivary glands combined with RNAi knockdown, small-molecule inhibitors and antibody-based blocking experiments is providing exciting results helping to dissect the signalling pathways that are involved in the morphogenetic events happening during salivary gland development.

Table 1. Salivary gland defects in mouse mutants

Salivary gland defects	Genotype	Gene product	Reference (first author)
Submandibular/sublingual gland			
Agenesis – initiation, no branching	*FGF10–/–*	Fibroblast growth factor 10	Ohuchi, 2000 [42], Jaskoll, 2005 [43]
	FGFR2-IIIb–/–	Fibroblast growth factor receptor 2 (IIIb)	De Moerlooze, 2000 [44], Revest, 2001 [45], Jaskoll, 2005 [43]
	$FGF8^{C/N}$; AP2α-IRESCre/+	Fibroblast growth factor 8	Jaskoll, 2004 [46]
Agenesis	*p63*	p63 (homolog to p53 tumour suppressor)	Yang, 1999 [47]
	Ptx1	Pitx1 – transcription factor	Szeto, 1999 [48]
Agenesis or hypoplasia	*Tbx1–/–*	Tbx1 – transcription factor of T-box family	Jerome, 2001 [49]
Agenesis, hypoplasia or wild-type correlating with external craniofacial phenotype	*Twsg1–/–*	Twisted gastrulation – secreted protein	Melnick, 2006 [50]
Severe hypoplasia	*Shh–/–*	Sonic hedgehog – signalling molecule	Jaskoll, 2004 [51]
Hypoplasia	*FGF10+/–*	Fibroblast growth factor 10	Entesarian, 2005 Jaskoll, 2005 [43]
	FGFR2-IIIb+/–	Fibroblast growth factor receptor 2 (IIIb)	Jaskoll, 2005 [43]
	FGFR2-IIIc+/–	Fibroblast growth factor receptor 2 (IIIc)	Jaskoll, 2002 [20]
	Itga3–/–; Itga6–/–	Integrin α_3 and Integrin α_6	De Arcangelis, 1999 [53], Rebustini, 2007 [54]
	MMP14	Matrix metalloprotease 14	Oblander, 2005 [55]
	Scrb1–/–	Scribbled 1 – cytoplasmic protein	Murdoch, 2003 [56]
	Six1–/–	Six1 – transcription factor	Laclef, 2003, McCoy 2009 [58]
Hypoplasia + disorganized epithelium + reduced lumen formation	*Lama5–/–*	Laminin α_5	Rebustini, 2007 [54]
Hypoplasia + absence of lumen	*$FGF8^{H/N}$ (hypomorph)*	Fibroblast growth factor 8	Jaskoll, 2004 [46]

Table 1. Continued

Salivary gland defects	Genotype	Gene product	Reference (first author)
Hypoplasia + decrease in granular convoluted ducts and acini	*Eda^{Ta}/Y*	Ectodysplasin-A	Blecher, 1983 [59], Jaskoll, 2003 [60]
Hypoplasia + disorganized mesenchyme	*Pax6*	Transcription factor	Jaskoll, 1999 [20]
Disorganized mesenchyme + reduced branching and lumen formation	*Bmp7–/–*	Bone morphogenetic protein 7	Jaskoll, 2002 [20]
Impaired growth, branching and maturation of the epithelium	*EGFR–/–*	Epidermal growth factor receptor	Jaskoll, 1999 [61], Haara, 2009 [62]
Severe dysplasia – absence of duct and acini	*Edar^{dl}*	Ectodysplasin-A receptor	Jaskoll, 2003 [60]
Severe dysplasia – absence of acini, lumen occlusion and disorganized epithelium	*MMTV-Cre; p120fl/fl*	p120 catenin	Davies, 2006 [63]
Disorganized basement membrane + abnormal cytodifferentiation	*Itga3b1–/–*	Integrin $\alpha_3\beta_1$	Menko, [64]
Shortened main ducts	*RARα$_1$; RARγ*	Retinoic acid receptors	Lohnes, 1994 [65]
Defect in duct maturation	*Cp2l1–/–*	Grainyhead-related transcription factor	Yamaguchi, 2006 [66]
Parotid gland			
Hypoplasia	*FGF10+/–*	Fibroblast growth factor 10	Entesarian, 2005 [52]
Defects specific to sublingual gland			
Agenesis	*Lama5–/–*	Laminin α_5	Rebustini, 2007 [54]
Dysplasia – cystic epithelial formations in the parenchyma	*RARα$_1$; RARγ*	Retinoic acid receptors	Lohnes, 1994 [65]
Retarded maturation of mucous-secreting acinar cells + disruption of cellular architecture	*Nkx2-3–/–*	Transcription factor	Biben, 2002 [67]
Shortened main duct	*RARβ$_2$; RARγ*	Retinoic acid receptors	Lohnes, 1994 [65]
Minor salivary glands			
Smaller glands with narrow ducts	*Nkx3.1–/–*	Transcription factor	Schneider, 2000 [68]

In addition to the culture of whole salivary gland rudiments, it is also possible, in early developing salivary glands, to cleanly separate the epithelium from the mesenchyme by enzymatic treatment and culture the epithelium alone either in growth factor-reduced matrigel [18] or in a laminin matrix [19]. In vitro culture of isolated salivary epithelium allows the researcher to study the effects of purified molecules on epithelial cell behaviour, in the absence of effects on the mesenchyme and from signals from the mesenchyme.

Early Embryonic Development of Salivary Glands: Initial Bud Formation

Although there is little doubt that the parotid gland derives from ectodermal tissue since it arises from the stomodeum, contradictory reports can be found in the literature regarding the origin of the sublingual and submandibular glands as to whether they are ectodermal or endodermal derivatives. In absence of a general endodermal marker, it is difficult to conclude either way. However, the endoderm is clearly capable of supporting salivary gland development given that minor salivary glands developing in the tongue, including von Ebner's glands, are undoubtedly of endodermal origin, as the epithelial layer covering the tongue is derived from the endoderm. Despite this controversy, major salivary glands are widely regarded as ectodermal organs, together with other exocrine glands such as mammary, sweat and sebaceous glands, and organs such as teeth, hair, scales, feathers and nails. All ectodermal organs originate from two adjacent tissues of distinct embryonic origin, the epithelium and the mesenchyme. Development of ectodermal organs proceeds through constant, sequential and reciprocal interactions between these two tissues translated at the molecular level by signalling molecules, which regulate cell proliferation, movements and differentiation. Although ectodermal organs exhibit great diversity in their appearance, molecular mechanisms involved in their development are strikingly similar, so what we learn about the development of salivary glands can help to reveal more general principles and conversely, mechanisms unravelled in other ectodermal organs can aid us to understand salivary gland development.

The mouse SMG develops through a series of interactions between the oral epithelium covering the first branchial arch, and a population of mesenchymal cells derived from the cranial neural crest, a migratory cell population that detaches from the embryonic neural epithelium [20]. The epithelial compartment will eventually give rise to the secretory endpieces of the salivary glands, the extensive network of ducts leading the salivary secretions into the oral cavity, and the myoepithelial cells. The mesenchymal compartment will produce the capsule surrounding the gland. Mouse SMG embryonic development is classically described in stages [21]. The first morphological sign of SMG development is observed around embryonic day 11.5 (E11.5). An epithelial thickening appears in the floor of the mouth, at the back of the first mandibular molar, adjacent to the developing tongue (fig. 2a, 1). This thickening develops at the bottom of the alveolo-lingual sulcus, a gutter-like groove that forms in the floor of the mouth as the result of the upward growth of the tongue rudiment (fig. 2b, top panel). This early stage is known as the prebud stage. SLG development starts one day later at E12.5 by a thickening of the oral epithelium located right next to the SMG on the buccal side (fig.

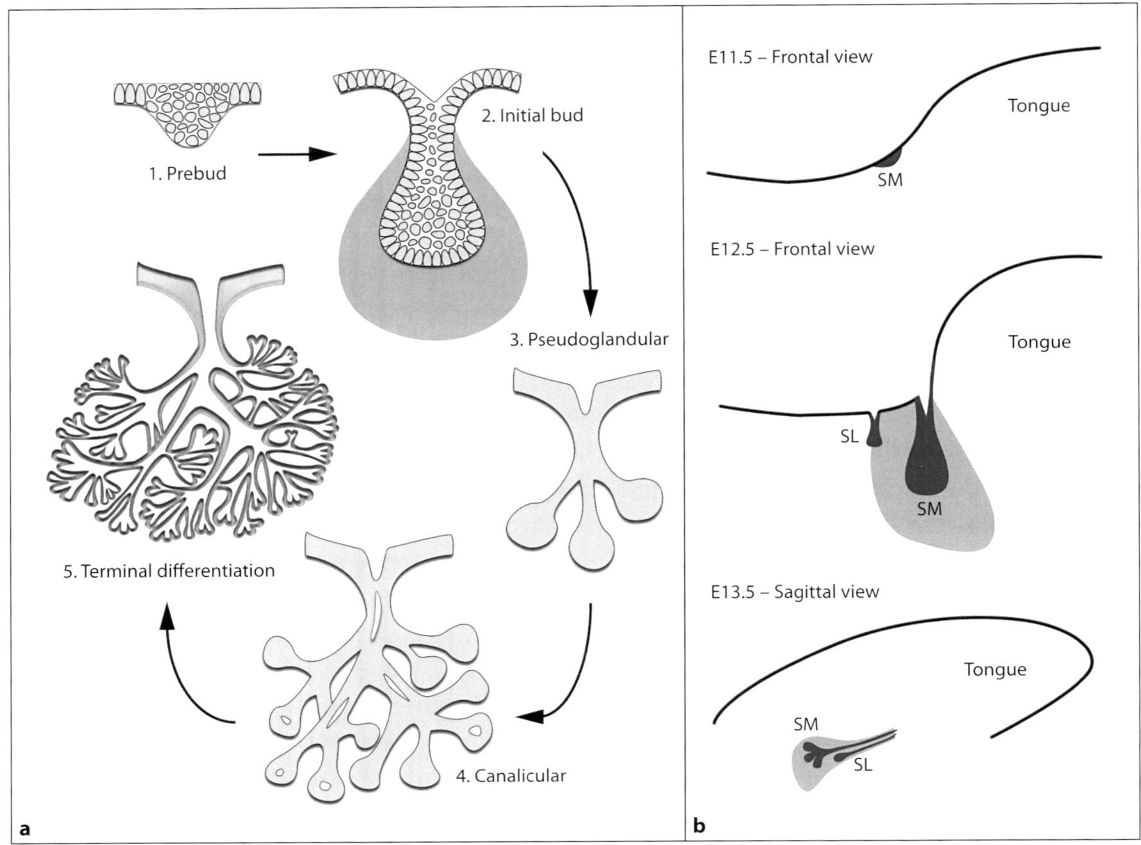

Fig. 2. Embryonic development of the submandibular (SM) and sublingual (SL) glands in the mouse. **a** Developmental stages of the SM gland. **b** Development of the SL gland relative to the SM gland. Condensed mesenchyme is indicated in dark grey around salivary gland epithelial rudiments in **a** and **b**.

2b, central panel). Importantly, both the SMG and SLG do not start developing in the location where their major excretory ducts open in the adult mouth. At E12.5, the SMG epithelial thickening invaginates in the underlying mesenchyme of the first branchial arch. Sustained epithelial proliferation in a downward direction leads to the formation of a thick solid epithelial stalk terminated by a bulge constituting the initial bud stage of SMG development (fig. 2a, 2). Concomitant to this process, mesenchymal cells condense around the SMG primordium. This well-defined mass of connective tissue represents the capsular rudiment of the gland. The initial bud, surrounded by condensed mesenchyme, will form the parenchyma of the SMG, whereas the main excretory duct of this gland is formed by closure, in a rostral direction, of the alveolo-lingual sulcus [22]. The SMG main excretory duct, as well as the one of the SLG, opens in the floor of the mouth at the sublingual caruncle, a mucosal fold located behind the lower incisors.

The SLG reaches an initial bud stage one day later than the SMG, at E13.5 (fig. 2b, bottom panel). The SLG epithelial bud penetrates the capsule rudiment of the SMG, rather than forming a separate mesenchymal condensation [15] (fig. 2b, central panel). On early day 13, the SMG reaches the late initial bud stage after an increase in volume of both the epithelial and mesenchymal compartments. Salivary gland rudiments can be cultured in vitro from an initial bud stage (E12), however most in vitro studies are carried out on late bud stage rudiments, since they develop better in culture.

A thick layer of extracellular matrix (ECM) separates the epithelial and mesenchymal compartments of the salivary gland primordium. This layer, called basement membrane, is secreted partly by epithelial cells, partly by surrounding mesenchymal cells [23] and amongst other functions primarily serves to anchor the epithelium to the underlying connective tissue. ECM is not only found in the basement membrane, but also in the interstitial matrix between cells of the connective tissue of the developing salivary gland. Sequeira et al. [pp. 48–77] offer a comprehensive review of the components of the ECM and their specific functions during salivary gland morphogenesis and differentiation. Interestingly, in the SMG initial bud, outer epithelial cells making contact with the basement membrane appear morphologically distinct from more central epithelial cells. Both the stalk and the bulb of the epithelial bud are composed of a regular outer layer of columnar cells surrounding a group of loosely arranged cells of irregular shape [15, 24] (fig. 2a, 2). This regular outer layer of cylindrical cells appears to be maintained throughout the embryogenesis of the mouse SMG. Whereas the inner cells divide actively, mitoses appear to be very rare in the outer layer [15]. Importantly, B1-immunoreactive proteins, which are a group of three proteins, SMGA, SMGB1 and SMGB2, secreted in acinar-cell progenitors of neonatal major salivary glands in mice and rats, are specifically present in these outer cells from a late bud stage onwards [24]. Although no definitive marker of acinar cell progenitors has been identified yet, this pattern of B1 protein distribution suggests that cells of secretory endpieces may originate from the cells lining the periphery of the epithelial rudiment and that epithelial cell commitment to an acinar fate may be determined very early during embryogenesis. Another indication that these outer epithelial cells may be committed early on during salivary gland development to a cell fate distinct from central epithelial cells is that at E13.5 (late bud stage) these cells do not express the same genes as epithelial cells that do not make contact with the basement membrane [personal observation] (fig. 3).

One central question about early development of salivary glands is what determines the sites where salivary glands will form. While this question has been answered for *Drosophila* larval salivary glands [25], a topic covered by Pirraglia and Myat [pp. 32–47], it remains an open question in mice owing to the lack of mouse mutants either developing ectopic supernumerary major salivary glands or forming known major salivary glands in abnormal locations. Likewise, transcription factors exhibiting an expression pattern restricted to the presumptive sites of salivary gland development have yet to be identified.

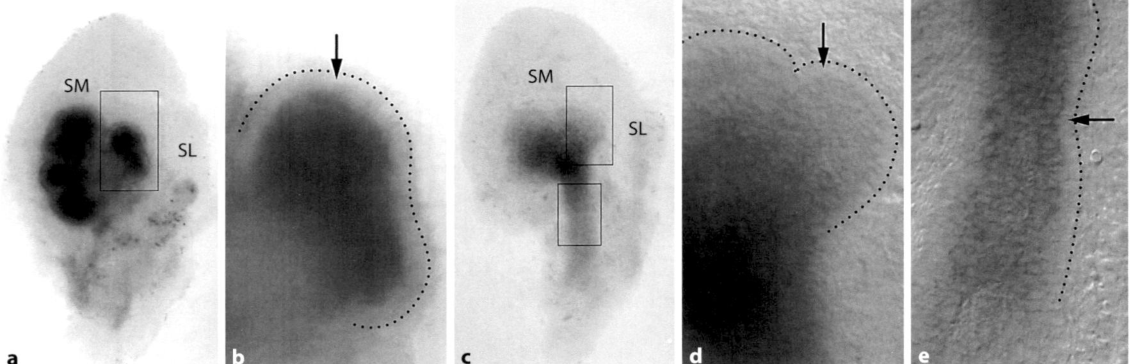

Fig. 3. At E13.5 (pseudoglandular stage), outer and inner epithelial cells of the developing submandibular (SM) gland do not express the same set of genes. The outline of the developing SM epithelium is indicated by a black dotted line in **b**, **d**, and **e.** Gene expression is visualized as dark staining. **a** Whole-mount in situ hybridization (WISH) for *cyclin D1*, which encodes a protein playing a role in cell proliferation through activation of cyclin-dependent kinases. *Cyclin D1* RNA is present in the epithelial end buds of the developing SM gland. **b** Higher magnification of the area boxed in **a**. Outer columnar epithelial cells (black arrow) are devoid of *cyclin D1* expression. **c** WISH for *Barx2*, which encodes a member of the homeobox transcription factor family. *Barx2* is expressed throughout the whole epithelium of the developing SM gland, both in the end buds and stalks. **d** Higher magnification of the bud area boxed in **c**. *Barx2* expression is absent in the outer epithelial cells (black arrow) that make contact with the basement membrane. **e** Higher magnification of the presumptive duct area boxed in **c**. *Barx2* is equally not expressed in the outer epithelial cells of the presumptive main duct of the SM gland (black arrow). SL = Sublingual gland.

Forming Branches to Escape Space Constraints

Between E13 and E14.5, a series of morphogenetic changes, collectively referred to as branching morphogenesis, occurs in the developing salivary gland turning an initial single epithelial bud into an array of epithelial branches that will eventually differentiate into a network of ducts, each terminating in a secretory endpiece. Branching is a strategy adopted by a group of cells, here the epithelium, to dramatically expand its surface area inside a confined space, providing a way to escape the space constraints created by the size of an organ. As such, it is a common developmental mechanism adopted by many organs in most animals. These organs, known as 'branching organs', include in humans and other placental mammals the salivary glands, mammary glands, kidneys, lungs, pancreas and prostate. Although the branched epithelia of these organs arise from different germ layers (endoderm for lungs and pancreas, mesoderm for kidneys, ectoderm for salivary glands and mammary glands) as branching morphogenesis starts to be elucidated in these organs, there is compelling evidence that they share a conserved set of molecular mechanisms promoting and regulating epithelial branching [26, 27].

The period during which branching morphogenesis occurs in salivary glands is called the pseudoglandular stage. During this phase, small invaginations or clefts form in the distal part of the initial epithelial bud (or end bud), which deepen and separate the bud into usually two or three parts (fig. 2a, 3). This parting process establishes branch points. How branch points are determined is currently unknown. End bud clefting is followed by outgrowth of epithelial branches (also called stalks) and new subsequent cleft formation in newly formed distal buds. Eventually, reiteration of this process of branch point formation and duct elongation leads to the formation of an increasingly larger and complex epithelial tree. During this process, mesenchymal cells that had condensed around the initial epithelial bud become looser, separated by large amounts of ECM composed of collagens, fibronectin and numerous other proteins [28, 29]. Branching is a dynamic process relying on coordinated cell proliferation, migration, and differentiation. Since the epithelium rapidly expands during salivary gland morphogenesis, it is tempting to see cell proliferation as the driving force behind epithelial branching. However, it does not seem to be the case, since cleft formation can occur in the absence of cell proliferation [30, 31].

Branching morphogenesis requires specific components of the ECM as well as soluble factors including growth factors and other signalling molecules as reviewed by Sequeira et al. [pp. 48–77]. Salivary gland epithelium clearly branches in response to signals from the surrounding mesenchymal tissue though we do not yet know of a transcription factor or secreted molecule that would render any epithelial tissue capable of undergoing branching morphogenesis. The instructive role of the mesenchyme during branching morphogenesis has been established by classic tissue recombination experiments. In this assay, epithelial and mesenchymal tissues of a developing salivary gland are separated and each tissue type is re-associated with an equivalent tissue isolated from another organ. The subsequent hybrid tissue is then cultured in vitro and assessed for developmental characteristics (such as the formation of epithelial branches, the kind of branching pattern or the type of product secreted) that provide information on the effect of one tissue upon another. Tissue recombinations performed with quail salivary gland tissues have produced very interesting results. Quails exhibit two types of salivary glands: an elongating-type, the anterior submaxillary gland, and a branching-type, the anterior lingual salivary gland. In vitro recombinations between tissues of these two glands showed that when elongating-type epithelium was associated with branching-type mesenchyme, the resulting salivary gland branched. Conversely, when elongating-type mesenchyme was recombined with branching-type epithelium, it caused the epithelium to elongate rather than develop branches [32]. Influence of the mesenchyme over the decision of the epithelium to branch or not to branch was further emphasized by heterotypic recombination between quail anterior submaxillary epithelium and mesenchyme from mouse SMG (a branching-type gland) leading to a branched gland. Interestingly, different branching organs form branches with different branch patterns that are defined by the length, diameter, shape and spacing of the branches. The mesenchyme also appears to regulate the branching pattern of the epithelium. For example, salivary gland mesenchyme can instruct either mammary or pituitary epithelium to form branches with patterns specific to the salivary glands [33, 34]. Although the stroma of a developing salivary gland is essential to trigger branching events and control branching patterns, it is the epithelium that seems to carry the information related to terminal cell differentiation, responsible for the functional specificity of secretory cells. While SMG mesenchyme recombined with mammary epithelium leads to a salivary branching pattern, the graft retains the ability to form milk-producing alveoli [33]. Likewise, in heterotypic recombinants of pancreatic epithelium with salivary mesenchyme in mouse [35] and parotid epithelium with submandibular mesenchyme and vice versa in rat [36], epithelial type-specific proteins are produced although the morphology of the epithelial structures is reminiscent of the organ providing the mesenchyme. However, the age of the epithelium in these grafts appears to be of highest importance since E8.5 pituitary epithelium recombined with E14 SMG mesenchyme can be respecified in α-amylase-secreting salivary gland epithelium while E12 pituitary epithelium cannot [34]. The distinct instructive roles of epithelial and mesenchymal salivary gland cells are reflected in the fact that branching morphogenesis and cytodifferentiation can be uncoupled. In early branching rat salivary glands deprived of most of their mesenchymal capsule, branching morphogenesis is impaired whereas cytodifferentiation proceeds normally [37]. It is important to note that from early stages of embryonic development, other branching structures are associated with the developing SMG such as nerves and blood vessels [15, 38]. In particular, the parasympathetic neuron network appears to follow the branching pattern of the salivary epithelium [38, 39], suggesting possible interactions between different branching structures in the same developing organ.

Turning Solid Epithelial Chords into Hollow Tubes

Unlike the initial buds forming at the beginning of lung or kidney development, which arise from the bending of a polarized epithelium leading to the formation of a hollow tube as the epithelium curves inwards [40], the SMG initial bud, similar to the one of the mammary gland, starts out as a solid chord of unpolarized cells. This solid initial bud then develops into a network of solid stalks and end buds, therefore requiring de novo formation of lumen to create epithelial tubes with a central empty space that will collect and lead salivary secretions into the oral cavity. Branches and buds hollow out in their centre by a process of cavitation during the canalicular (E15.5) (fig. 2a, 4) and terminal bud (E18.5) (fig. 2a, 5) stages. This process involves apoptosis (programmed cell death) of centrally located epithelial cells in the epithelial stalks, which basically are the cells that do not make contact with the basement membrane. Small separate lumina are formed in several places of the epithelial rudiment (fig. 2a, 4), which later coalesce to eventually form a continuous lumen (fig. 2a, 5). During the terminal bud stage, lumina are found both in the presumptive ducts and secretory endpieces. It is only in the late phase of this stage that they become continuous. Concomitant to the event of epithelial cavitation, dramatic cellular changes occur within the epithelial cells at the periphery of the stalks, which are the cells in direct contact with the basement membrane and ECM. These outside epithelial cells become polarized with a basal surface facing out, towards the basement membrane and surrounding tissues and an apical surface facing the presumptive lumen. Simultaneously, these cells also become tightly anchored to each other via specialized junctions on their lateral membranes. Although studies in the mammary gland have shown that apicobasal polarization precedes apoptosis [41], the timing of these events respective to each other has not yet been studied during mammalian salivary gland development. Strengthening of cell contacts via epithelial junctions in the peripheral epithelial cells achieves two important ductal functions. Firstly, this epithelial layer becomes impermeable, preventing passive leaks in or out of the newly created tubes in between the cells of the epithelium (paracellular transport). Secondly, junctions create compartments within the cytoplasmic membrane enabling specialized functions (such as for example secretion or absorption) to take place on different cell surfaces. Signalling pathways and molecules involved in lumen formation are covered in detail by Wells and Patel [pp. 78–89]. Interestingly, cell death may not be the unique mechanism used to create tubes in the SMG. Indeed, formation of the main excretory ducts, which has yet to be studied, has been described to occur by closure of a groove in the epithelium of the floor of the mouth [22]. Whether it involves a wrapping mechanism, as observed during primary neurulation in vertebrates, or a budding process with a downward extension of an epithelial invagination, remains to be investigated.

Conclusions and Perspectives

Specialized cell types (serous or mucous cells) appear in secretory endpieces during the terminal differentiation stage of salivary gland embryonic development, although mouse and rat salivary glands are differentiated predominantly postnatally and reach complete maturation in the weeks following birth. The constant interactions between different cell types that are crucial during salivary gland morphogenesis and differentiation do not cease once the gland has reached its final size and full maturation. Similar to other organs, salivary glands have to be maintained throughout the life of one individual and there is therefore a constant turnover of cells in these organs. There is currently a growing interest in the question of whether adult salivary glands contain multipotent stem cells capable of generating all of the different salivary cell types. The ultimate goal of this area of research is to regenerate salivary gland tissues of patients with irreversible salivary hypofunctions, who have a very poor quality of life. This subject is covered by Carpenter and Cotroneo [pp. 107–128]. Lombaert and Hoffman [pp. 90–106] also present an original work that looks for the first time at the presence of epithelial stem/progenitor cells in embryonic salivary glands at different stages of development. In conclusion, progress in the understanding of salivary gland development, differentiation and maintenance highlights how the study of a specific branched tubular organ helps unravelling common strategies and conserved molecular mechanisms shared by many vital human organs.

References

1 Milne-Edwards H: Leçons sur la physiologie et l'anatomie comparée de l'homme et des animaux/faites a la faculté des sciences de Paris. Paris, Librairie de Victor Masson, 1860, vol 6.

2 Born G: Observations anatomiques sur la grande Lamproie; in Audouin Dumas B (ed): Annales des sciences naturelles. Paris, Crochard, Libraire-Editeur, 1828, vol. 13.

3 Pedersen AM, Bardow A, Jensen SB, Nauntofte B: Saliva and gastrointestinal functions of taste, mastication, swallowing and digestion. Oral Dis 2002;8:117–129.

4 Zelles T, Purushotham KR, Macauley SP, et al: Saliva and growth factors: the fountain of youth resides in us all. J Dent Res 1995;74:1826–1832.

5 Groschl M: The physiological role of hormones in saliva. Bioessays 2009;31:843–852.

6 Gresik EW: The granular convoluted tubule cell of rodent submandibular glands. Microsc Res Tech 1994;27:1–24.

7 Frey R, Markgraf U, Hofmann RR: Evolutionary morphology of the zygomatic gland and lacrimal bulla in roe deer (*Capreolus capreolus* Linnaeus, 1758 – Mammalia, Cervidae). Zool Anz 2001;240:181–195.

8 Lacassagne A: Dimorphisme sexuel de la glande sous-maxillaire chez la souris. CR Soc Biol Paris 1940;133:180–181.

9 Pinkstaff CA: Salivary gland sexual dimorphism: a brief review. Eur J Morphol 1998;36(suppl):31–34.

10 Hosoi K, Kobayashi S, Ueha T: Sex difference in L-glutamine D-fructose-6-phosphate aminotransferase activity of mouse submandibular gland. Biochim Biophys Acta 1978;543:283–292.

11 Treister NS, Richards SM, Lombardi MJ, et al: Sex-related differences in gene expression in salivary glands of BALB/c mice. J Dent Res 2005;84:160–165.

12 Dawes C: Rhythms in salivary flow rate and composition. Int J Chronobiol 1974;2:253–279.

13 Segal K, Lisnnyansky I, Nageris B, Feinmesser R: Parasympathetic innervation of the salivary glands. Oper Tech Otolaryngol 1996;7:333–338.

14 Nanci A: Ten Cate's Oral Histology: Development, Structure, and Function, ed 6. St Louis, Mosby, 2003.

15 Borghese E: The development in vitro of the submandibular and sublingual glands of *Mus musculus*. J Anat 1950;84:287–302.

16 Tucker AS: Salivary gland development. Semin Cell Dev Biol 2007;18:237–244.

17 Larsen M, Wei C, Yamada KM: Cell and fibronectin dynamics during branching morphogenesis. J Cell Sci 2006;119:3376–3384.

18 Nogawa H, Takahashi Y: Substitution for mesenchyme by basement membrane-like substratum and epidermal growth factor in inducing branching morphogenesis of mouse salivary epithelium. Development 1991;112:855–861.

19 Steinberg Z, Myers C, Heim VM, et al: FGFR2b signaling regulates ex vivo submandibular gland epithelial cell proliferation and branching morphogenesis. Development 2005;132:1223–1234.

20 Jaskoll T, Zhou YM, Chai Y, et al: Embryonic submandibular gland morphogenesis: stage-specific protein localization of FGFs, BMPs, Pax6 and Pax9 in normal mice and abnormal SMG phenotypes in FgfR2-IIIc(+/δ), BMP7(–/–) and Pax6(–/–) mice. Cells Tissues Organs 2002;170:83–98.

21 Melnick M, Jaskoll T: Mouse submandibular gland morphogenesis: a paradigm for embryonic signal processing. Crit Rev Oral Biol Med 2000;11:199–215.

22 Hamilton WJ, Boyd JD, Mossman HW: Human Embryology: Prenatal Development of Form and Function. Cambridge, WHS, 1945.

23 Brownell AG, Bessem CC, Slavkin HC: Possible functions of mesenchyme cell-derived fibronectin during formation of basal lamina. Proc Natl Acad Sci USA 1981;78:3711–3715.

24 Walker JL, Menko AS, Khalil S, et al: Diverse roles of E-cadherin in the morphogenesis of the submandibular gland: insights into the formation of acinar and ductal structures. Dev Dyn 2008;237:3128–3141.

25 Andrew DJ, Henderson KD, Seshaiah P: Salivary gland development in *Drosophila melanogaster*. Mech Dev 2000;92:5–17.

26 Lu P, Werb Z: Patterning mechanisms of branched organs. Science 2008;322:1506–1509.

27 Davies JA: Do different branching epithelia use a conserved developmental mechanism? Bioessays 2002;24:937–948.

28 Macauley SP, Tarnuzzer RW, Schultz GS, et al: Extracellular matrix gene expression during mouse submandibular gland development. Arch Oral Biol 1997;42:443–454.

29 Hardman P, Spooner BS: Localization of extracellular matrix components in developing mouse salivary glands by confocal microscopy. Anat Rec 1992; 234:452–459.

30 Spooner BS, Bassett KE, Spooner BS Jr: Embryonic salivary gland epithelial branching activity is experimentally independent of epithelial expansion activity. Dev Biol 1989;133:569–575.

31 Nakanishi Y, Morita T, Nogawa H: Cell proliferation is not required for the initiation of early cleft formation in mouse embryonic submandibular epithelium in vitro. Development 1987;99:429–437.

32 Nogawa H, Mizuno T: Mesenchymal control over elongating and branching morphogenesis in salivary gland development. J Embryol Exp Morphol 1981;66:209–221.

33 Sakakura T, Nishizuka Y, Dawe CJ: Mesenchyme-dependent morphogenesis and epithelium-specific cytodifferentiation in mouse mammary gland. Science 1976;194:1439–1441.

34 Kusakabe M, Sakakura T, Sano M, Nishizuka Y: A pituitary-salivary mixed gland induced by tissue recombination of embryonic pituitary epithelium and embryonic submandibular gland mesenchyme in mice. Dev Biol 1985;110:382–391.

35 Rutte, WJ, Wessells NK, Grobstein C: Control of specific synthesis in the developing pancreas. Natl Cancer Inst Monogr 1964;13:51–65.

36 Lawson KA: The role of mesenchyme in the morphogenesis and functional differentiation of rat salivary epithelium. J Embryol Exp Morphol 1972;27: 497–513.

37 Cutler LS: The dependent and independent relationships between cytodifferentiation and morphogenesis in developing salivary gland secretory cells. Anat Rec 1980;196:341–347.

38 Patel VN, Rebustini IT, Hoffman MP: Salivary gland branching morphogenesis. Differentiation 2006;74: 349–364.

39 Coughlin MD: Early development of parasympathetic nerves in the mouse submandibular gland. Dev Biol 1975;43:123–139.

40 Hogan BL, Kolodziej PA: Organogenesis: molecular mechanisms of tubulogenesis. Nat Rev Genet 2002; 3:513–523.

41 Debnath J, Mills KR, Collins NL, et al: The role of apoptosis in creating and maintaining luminal space within normal and oncogene-expressing mammary acini. Cell 2002;111:29–40.

42 Ohuchi H, Hori Y, Yamasaki M, et al: FGF10 acts as a major ligand for FGF receptor 2 IIIb in mouse multi-organ development. Biochem Biophys Res Commun 2000;277:643–649.

43 Jaskoll T, Abichaker G, Witcher D, et al: FGF10/ FGFR2b signaling plays essential roles during in vivo embryonic submandibular salivary gland morphogenesis. BMC Dev Biol 2005;5:11.

44 De Moerlooze L, Spencer-Dene B, Revest JM, et al: An important role for the IIIb isoform of fibroblast growth factor receptor 2 in mesenchymal-epithelial signalling during mouse organogenesis. Development 2000;127:483–492.

45 Revest JM, Spencer-Dene B, Kerr K, et al: Fibroblast growth factor receptor 2-IIIb acts upstream of Shh and Fgf4 and is required for limb bud maintenance but not for the induction of Fgf8, Fgf10, Msx1, or Bmp4. Dev Biol 2001;231:47–62.

46 Jaskoll T, Abichaker G, Witcher D, et al: FGF8 dose-dependent regulation of embryonic submandibular salivary gland morphogenesis. Dev Biol 2004;268: 457–469.

47 Yang A, Schweitzer R, Sun D, et al: p63 is essential for regenerative proliferation in limb, craniofacial and epithelial development. Nature 1999;398:714–718.

48 Szeto DP, Rodriguez-Esteban C, Ryan AK, et al: Role of the bicoid-related homeodomain factor Pitx1 in specifying hindlimb morphogenesis and pituitary development. Genes Dev 1999;13:484–494.

49 Jerome LA, Papaioannou VE; DiGeorge syndrome phenotype in mice mutant for the T-box gene, Tbx1. Nat Genet 2001;27:286–291.

50 Melnick M, Petryk A, Abichaker G, et al: Embryonic salivary gland dysmorphogenesis in twisted gastrulation-deficient mice. Arch Oral Biol 2006;51:433–438.

51 Jaskoll T, Leo T, Witcher D, et al: Sonic hedgehog signaling plays an essential role during embryonic salivary gland epithelial branching morphogenesis. Dev Dyn 2004;229:722–732.

52 Entesarian M, Matsson H, Klar J, et al: Mutations in the gene encoding fibroblast growth factor 10 are associated with aplasia of lacrimal and salivary glands. Nat Genet 2005;37:125–127.

53 De Arcangelis A, Neuville P, Boukamel R, et al: Inhibition of laminin α_1-chain expression leads to alteration of basement membrane assembly and cell differentiation. J Cell Biol 1996;133:417–430.

54 Rebustini IT, Patel VN, Stewart JS, et al: Laminin α_5 is necessary for submandibular gland epithelial morphogenesis and influences FGFR expression through β_1 integrin signaling. Dev Biol 2007;308:15–29.

55 Oblander SA, Zhou Z, Gálvez BG, et al: Distinctive functions of membrane type 1 matrix-metalloprotease (MT1-MMP or MMP-14) in lung and submandibular gland development are independent of its role in pro-MMP-2 activation. Dev Biol 2005;277:255–269.

56 Murdoch JN, Henderson DJ, Doudney K, et al: Disruption of scribble (Scrb1) causes severe neural tube defects in the circle tail mouse. Hum Mol Genet 2003;12:87–98.

57 Laclef C, Souil E, Demignon J, Maire P: Thymus, kidney and craniofacial abnormalities in Six1-deficient mice. Mech Dev 2003;120:669–679.

58 McCoy EL, Kawakami K, Ford HL, Coletta RD: Expression of Six1 homeobox gene during development of the mouse submandibular salivary gland. Oral Dis 2009;15:407–413.

59 Blecher SR, Debertin M, Murphy JS: Pleiotropic effect of Tabby gene on epidermal growth factor-containing cells of mouse submandibular gland. Anat Rec 1983;207:25–29.

60 Jaskoll T, Zhou YM, Trump G, Melnick M: Ectodysplasin receptor-mediated signaling is essential for embryonic submandibular salivary gland development. Anat Rec A Discov Mol Cell Evol Biol 2003;271:322–331.

61 Jaskoll T, Melnick M: Submandibular gland morphogenesis: stage-specific expression of TGF-α/EGF, IGF, TGF-β, TNF, and IL-6 signal transduction in normal embryonic mice and the phenotypic effects of TGF-β_2, TGF-β_3, and EGF-r null mutations. Anat Rec 1999;256:252–268.

62 Haara O, Koivisto T, Miettinen PJ: EGF-receptor regulates salivary gland branching morphogenesis by supporting proliferation and maturation of epithelial cells and survival of mesenchymal cells. Differentiation 2009;77:298–306.

63 Davis MA, Reynolds AB: Blocked acinar development, E-cadherin reduction, and intraepithelial neoplasia upon ablation of p120-catenin in the mouse salivary gland. Dev Cell 2006;10:21–31.

64 Menko AS, Kreidberg JA, Ryan TT, et al: Loss of $\alpha_3\beta_1$ integrin function results in an altered differentiation program in the mouse submandibular gland. Dev Dyn 2001;220:337–349.

65 Lohnes D, Mark M, Mendelsohn C, et al: Function of the retinoic acid receptors (RARs) during development (I). Craniofacial and skeletal abnormalities in RAR double mutants. Development 1994;120:2723–2748.

66 Yamaguchi Y, Yonemura S, Takada S: Grainyhead-related transcription factor is required for duct maturation in the salivary gland and the kidney of the mouse. Development 2006;133:4737–4748.

67 Biben C, Wang WW, Harvey RP: NK-2 class homeobox genes and pharyngeal/oral patterning: Nkx2-3 is required for salivary gland and tooth morphogenesis. Int J Dev Biol 2002;46:415–422.

68 Schneider A, Brand T, Zweigerdt R, Arnold H: Targeted disruption of the Nkx3.1 gene in mice results in morphogenetic defects of minor salivary glands: parallels to glandular duct morphogenesis in prostate. Mech Dev 2000;95:163–174.

Isabelle Miletich
Department of Craniofacial Development and Orthodontics
Floor 27, Guy's Hospital, London SE1 9RT (UK)
Tel. +44 207 188 8038, Fax +44 207 188 1674
E-Mail isabelle.miletich@kcl.ac.uk

Tucker AS, Miletich I (eds): Salivary Glands. Development, Adaptations and Disease.
Front Oral Biol. Basel, Karger, 2010, vol 14, pp 21–31

Salivary Gland Adaptations: Modification of the Glands for Novel Uses

Abigail S. Tucker

Department of Craniofacial Development and Orthodontics, Guy's Hospital, London, UK

Abstract

Salivary glands across the animal kingdom show a huge array of shapes, sizes and variations in number within the oral cavity. Some are branched, others elongated and unbranched, some are small and numerous, while others are greatly enlarged. In most species the salivary glands are used to produce saliva, which lubricates the oral cavity, aids digestion and protects the oral mucosa and dentition. In some species, however, the glands have been modified so that they function in new and diverse ways. Examples include the ability to create thread from the salivary glands to create hides and cocoons, as shown by some arthropods, the ability to create nests out of saliva, as shown by the *Aerodramus* swiftlets, or the ability to produce venom from modified salivary glands. In this chapter these different adaptations will be discussed, looking at how the salivary glands have become adapted to their new role, with insights from developmental biology and evolution.

Salivary glands come in all shapes and sizes. Often the variation is linked to differences in diet. For example it has been reported that ruminant species that eat more grass have smaller salivary glands, reflecting a reduced requirement for salivary tannin-binding proteins [1]. Bats also show large variations in the secretion capacity of the striated ducts of their glands depending on whether they are fruit or insect eating [2]. In bird species that eat small insects or seeds the salivary glands are well developed, while some fish-eating birds such as pelicans have been reported to have no salivary glands [3, 4]. In some species, however, the salivary glands have been modified so that they function in new and often unusual ways. A few such adaptations have been picked in this chapter for further discussion – production of silk, nest-building saliva, and venomous saliva.

Silk Production

The classic example of arthropod silk production from salivary glands is the mulberry silkworm, *Bombyx mori.* The *Bombix* caterpillar creates a silk cocoon from its salivary glands and is farmed to produce silk. The silkworm encloses itself in approximately 1 mile of silk filament, a liquid that solidifies on contact with air. The silk glands consist of a pair of tubular structures arising from the labial segment and extending backwards through the body. The two glands converge at the anterior and form a common secretory duct, similar to salivary glands in *Drosophila* [5]. In addition, silkworm larvae possess a distinct pair of salivary glands derived form the mandibular segment, which aid in digestion. The silk gland placodes grow and differentiate into an anterior, middle and posterior gland region. At the most anterior end the combined ducts open in the spinneret, which spins out the silk proteins to form threads. The silk structural proteins (fibroin L and H chains and fibrohexamerin) are synthesized in the posterior segment, while the middle segment synthesizes the glue proteins, sericin 1 and 2 [5]. The initiation of the silk glands has been shown to be dependent on the homeobox gene *sex combs reduced (Scr)*, as ectopic expression in the thorax leads to the induction of ectopic glands [6]. Silk glands share similarities with the *Drosophila* larval salivary glands, and have been long refereed to as homologous structures [7]. In keeping with this, expression of a *Bombyx* silk gene, *fibrohexamerin,* also known as *P25*, can be specifically driven in the anterior salivary glands of transgenic *Drosophila* larvae. Thus despite the fact that these two species diverged 240 million years ago, and that *Drosophila* does not produce silk, the mechanisms regulating transcription have been conserved between the salivary gland of the fruitfly and the silk gland of the silkmoth [8]. The fact that only the anterior region of the *Drosophila* salivary gland expressed *fibrohexamerin* indicates that it is this section that is homologous to the cells of the posterior silk gland [8]. In addition, *Scr* directs salivary gland formation in the labial segment in *Drosophila*, thus playing a central role in salivary and silk glands [9]. A detailed description of *Drosophila* salivary glands can be found in the following chapter by Pirraglia and Myat [pp. 32–47].

Silk production from adapted salivary glands is also observed in a number of other insects, such as ants, bees and wasps, although the arrangement of secreted proteins and the resulting macromolecular pattern is different [10].

Nest Formation

In the *Aerodramus* swifts the sublingual glands are enlarged and produce mucous-rich saliva that hardens when exposed to air. In two *Aerodramus* species (*Collocalia fuciphaga* and *C. germani*, the white-nest swiftlets) saliva alone forms the nest, and is used to make bird's nest soup. The nests of the swiftlets rank amongst the world's most expensive animal products. Due to increased demand for the edible nests, nest yields have been reported to have fallen by 83% over the last 10 years in the Andaman Islands, making the edible-nest swifts in need of conservation [11]. The saliva forms not only the nest but secures the bracket-shaped nest to the cave wall, and can cement the eggs to the nest. The swifts' sublingual glands were proposed to swell over the breeding season, as researchers found swollen glands in only those species that were in the mating season [12]. In keeping with this idea that the glands change on breeding, fledgling birds were found to have small non-secretory sublingual glands [13]. The increase in size of the sublingual gland in these birds therefore does not occur during embryonic development. Not all female swifts in breeding condition were found to have enlarged salivary glands, whereas all males had active sublingual glands. The presence or absence of the nest appeared to be the most important external factor controlling the activity of the glands [13]. Enlarged salivary glands are a feature of many swifts and the enlargement process has been followed in the chimney swift *(Chaetura pelagica)* [14]. At the height of the breeding season, each gland swells until it occupies completely the area between the floor of the month and the skin. When weighed, adult male glands ranged from 3.2 to 68 mg, a huge difference in size that correlates with cyclical gonadal enlargement. Each gland becomes highly vascularized and then regresses with a mass of cellular debris located in the lumens. This debris appears to represent cells which have been cast off from the tubules of the gland as it shrinks [14]. Such dramatic increases and decreases of the glands are very interesting and indicate that a population of stem cells might be active in the glands to aid in a form of regeneration [see chapters by Lombaert and Hoffman, pp. 90–106, and Thomas et al., pp. 129–146].

Venom Production

Across the animal kingdom, one striking adaptation of salivary glands is their use as a source of venom production. Venom produced by salivary glands has evolved in both vertebrates and invertebrates several times. These modified glands play an essential role in prey acquisition, prey immobilization and defence.

In many marine carnivorous gastropods, venoms such as tetramine and echotoxins have been identified in the salivary glands, while in the cephalopods, cephalotoxins and tetrodotoxin (TTx), the potent neurotoxin found in puffer fish, are produced by the posterior salivary glands [15]. Tetrodotoxin is produced by endosymbiotic bacteria, rather than being an endogenously produced toxin, and is not exclusively produced by the glands, with other regions of the body also forming the toxin [16]. The posterior glands are situated in the abdomen and are constructed of branched, convoluted tubules. The paired glands have short ducts that connect together via a long secretory duct that reaches up to the beak. In the octopus, in addition to the large posterior glands, anterior salivary glands are located closely associated with the buccal mass and account for most of the mucus secretion in the mouth [17]. The anterior glands, however, also express some toxins, and may play a role in prey immobilization [18].

Some of the toxins found in the cephalopods are related to those observed in reptiles and arthropods, indicating convergent protein recruitment [18]. The similarity

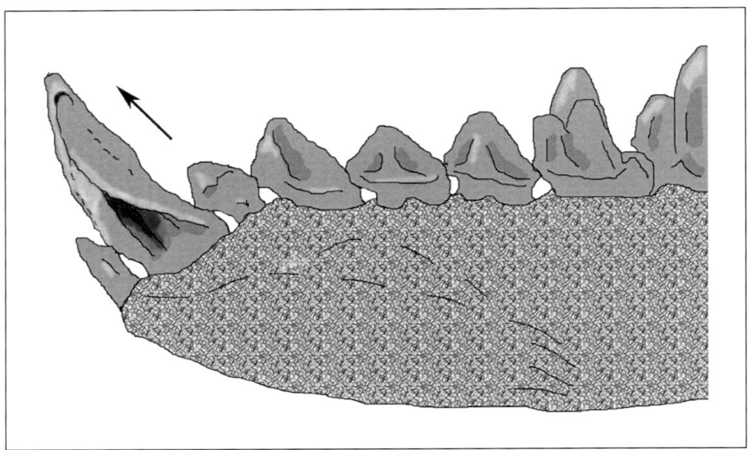

Fig. 1. Schematic of the lower jaw of a *Solenodon*. Arrow indicates flow of venomous saliva through the grooved lower incisor.

in the types of protein used indicates that restrictions are in place which limit which proteins are amenable for use as toxins [18].

Venomous Saliva in Mammals

In the vertebrates, venomous saliva is most commonly associated with reptiles, although examples have been documented in mammals. Some species of shrew, for example, have been shown to produce toxic compounds in their saliva. In the short-tailed shrew, *Blarina brevicauda*, the venom exerts a kallikrein-like proteolytic activity similar to that of lizard venom [19]. The modern shrew venomous saliva, however, is not associated with any adaptations to the teeth, although the concave-shaped lower incisors might play a role in channelling the venom towards the bite site. The Caribbean *Solenodon,* a primitive shrew-like insectivore, also has venomous saliva, but in contrast to the poisonous shrews, this adaptation is associated with an enlarged and grooved lower incisor [20] (fig. 1). Solenodons comprise two living species the Cuban and Haitian (or hispaniolan) Solenodon, *Solenodon cubanus and S. paradoxus*. Both species are highly endangered, with the Cuban Solenodon thought extinct until a specimen was found in 2003. In keeping with its modification for venom production, the *Solenodon's* submandibular gland is enlarged and releases venomous saliva into the grooved incisor [20]. Fossil evidence supports the evolution of such a venom delivery system in early mammals, with venom-conducting upper canines identified in fossils of *Bisonalveus brown*, a small eutherian classified in the order Cimolesta from around 60 million years [21]. A venom apparatus has also been documented in two different species of extinct 'giant' shrews from 1.25 to 1.5 million years [22]. In these soricine taxa grooves were found along the lower incisors, similar to those seen in modern-day *Solenodons.* A venom delivery system, with coordinated use of salivary glands and teeth, did therefore evolve in mammals, as it did in reptiles. It may be that the relatively large sizes of these extinct soricines, and the extant *Solenodon* (28–32 cm in length), led to pressure for capture of larger prey, and a more successful means of immobilizing prey through a coordinated venom delivery system. The *Solenodon*, for example, is not strictly an insectivore and will tackle chicken and rodents [23]. Given this it is interesting to question why the use of such a venom delivery system is so rare in extant mammals. One possibility is that modern mammals have adopted faster methods to subdue prey, using 'tooth and claw', rather than relying on the rather slower method of venom [23]. The fact that the Solenodon, the sole living genus with a venom delivery system, is almost extinct due to predation by larger mammals, stresses the fact that for mammals a venom delivery system does not actually provide an enormous advantage.

Evolution of the Venom System in Reptiles

In reptiles one of the most striking adaptations is the evolution of distinct poison glands. Snakes, such as the elapids (cobras) and vipers, have venom glands associated with complex upper jaw fangs that are hollowed out by a folding of the developing tooth germ to allow the venom to be injected into prey [24]. These fangs are positioned at the front of the jaw. In contrast in the colubridae, the grooved fangs are located at the back of the upper jaw and the associated venom gland is known as the Duvernoy's gland. Evolution of this venom delivery system is thought to have been a major player in allowing the massive radiation of advanced snakes. Recent research has shown that the front and back fangs develop from a similar region at the back of the upper jaw during embryonic development [25]. This work highlights the homology of the venom glands of front-fanged snakes and the Duvernoy's glands of the back-fanged snakes, and the use of a separate term for the two glands is unnecessary. The venom gland forms two distinct parts, the posterior venom gland and the more anterior accessory gland (fig. 2a, b). In the viperids these two parts of the gland have very different histology. The main gland is formed from repeatedly branching tubules arranged around a large central lumen in which the venom is stored, while the smaller accessory gland consists of mucus-secreting cells, and stains positive for alcian blue (fig. 2c, d).

The use of venom produced by modified salivary glands was initially thought to be restricted to advanced snakes and the helodermatid lizards, such as the beaded lizard and glia monster, indicating independent origin in these two lines. In the helodermatid lizards, the venom is not associated with glands of the upper jaw but of the lower jaw. Recent research, however, has shown that the use of venom is much more widespread with *Varanus* lizards, such as the Komodo dragon and iguania possessing toxic

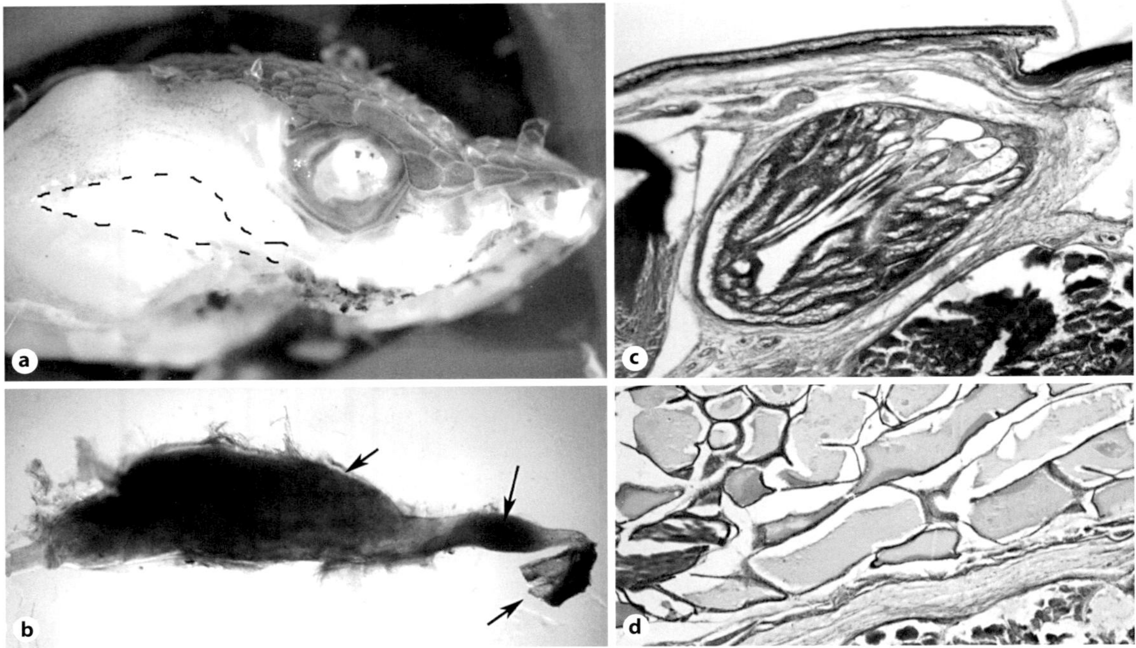

Fig. 2. Venom gland in the pit viper, *Trimeresurus alboloabris*. **a** Juvenile pit viper with scales dissected over the region of the venom gland (outlined by dashed lines). **b** Dissected venom gland showing the main venom gland that leads to the accessory gland before terminating in a splayed duct that opens next to the hollow fangs (arrows). **c** Section through accessory gland showing the dense mucous cells surrounding the central duct. **d** Section through venom gland showing the large vacuoles associated with this gland.

venom in their salivary secretions [26, 27]. In the Komodo dragon, the venom glands are encapsulated in a sheath of connective tissue and possess large distinct lumens, fed by an extensive network of ducts, providing combined 1 ml liquid storage volume [27]. That the Komodo dragon possesses venom puts the widespread belief that these giant lizards were able to debilitate their prey by the use of toxic bacteria-rich saliva in serious doubt. It is often cited that these lizards induce septicaemia in their prey after biting, and then track down their infected prey. In fact, these lizards were found to not have a particularly potent bacteria cocktail but are thought to induce shock by the use of venom [27]. So the utilization of bacteria in reptile saliva appears unlikely as a salivary gland associated adaptation.

In the iguania, venom was associated with glands in both the upper and lower jaw, indicating that this may have been the basic state, and that use of venom was then subsequently lost in the lower jaw in snakes, and in the upper jaw in other poisonous lizards. This new evidence puts the origin of venom delivery systems at around 200 million years ago, much earlier than previously suggested, and allows poisonous reptiles to be grouped into a new clade, the Toxicofera [26].

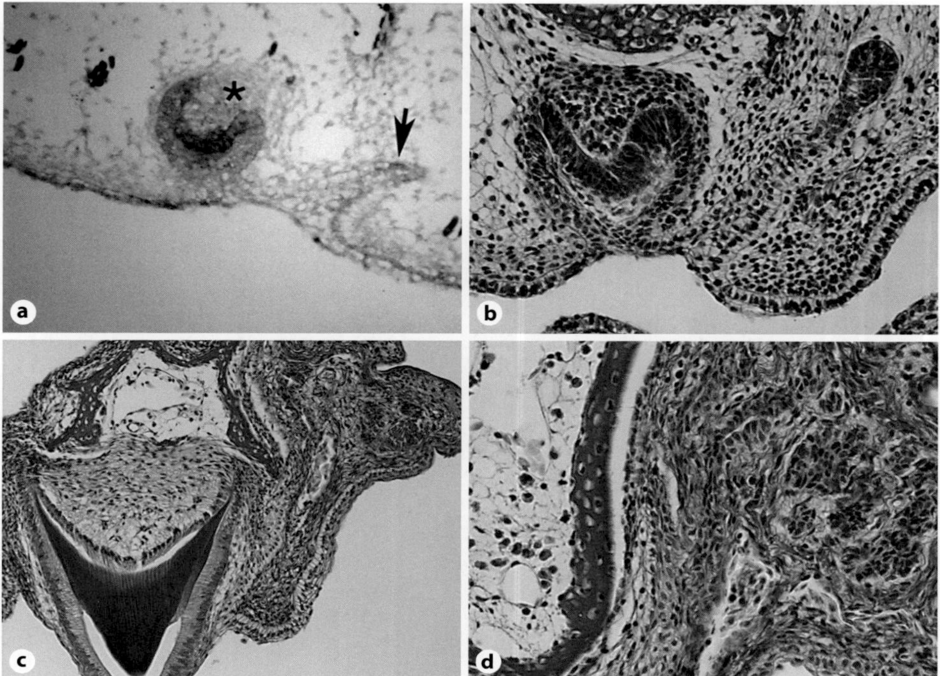

Fig. 3. Dental gland development in the chameleon. **a** Early gland and tooth development in a chameleon embryo. The tooth (*) and dental gland (arrow) develop side by side in the oral cavity. **b** As the tooth developments the associated dental gland forms an extended bud-like structure. **c** As the tooth starts to undergo differentiation, the dental gland shows a branched morphology with the formation of lumens. The main duct links the gland to the base of the tooth. **d** Close-up of branching dental gland.

Interactions of Glands and Teeth in the Reptilian Venom System

From the salivary gland point of view the evolution of the reptilian venom system can be thought of as occurring in two steps: the glands needed to be closely associated with the teeth, and the venom toxins needed to be produced.

Venom glands are modified dental glands. Dental glands are found in a number of reptiles. They are thought to function by lubricating the oral region around the teeth. Each tooth may be associated with a gland, and therefore there can be a large number of dental glands forming in the mouth [28]. The dental glands have a very interesting pattern of development that is intricately linked to tooth development. In many lizards, the dental lamina forms next to a second lamina-like structure found on the buccal side of the jaw [28] (fig. 3a). Whether these two lamina structures form from a single thickening is unclear, but is an intriguing possibility. As the tooth germs form from the dental lamina, the connected gland lamina forms a bud-like structure at the end of the gland lamina (fig. 3b). The dental gland develops buccal

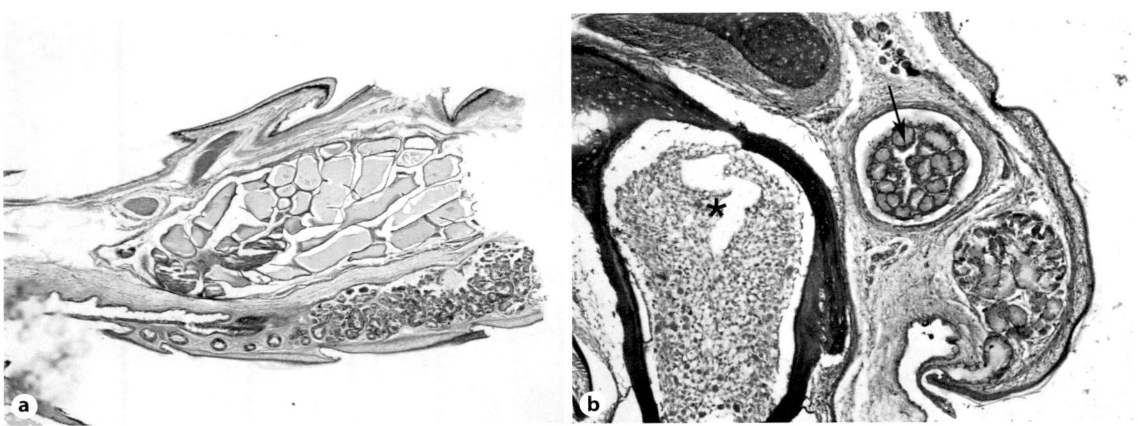

Fig. 4. Histology of the venom and salivary glands in *Trimeresurus*. **a** Sagittal section. The supralabial salivary gland runs underneath the venom gland and has a very different morphology, showing a much tighter branched morphology. **b** Frontal section. The duct of the supralabial gland opens into the oral cavity way from the developing tooth (*). The accessory gland of the venom gland is shown above (arrow).

to the supralabial salivary gland, which can be seen developing at this stage (fig. 4b). As tooth development progresses through the bell stage, the dental gland starts undergoing branching. At the late bell stage of tooth development, when cytodifferentiation has started to occur, the dental gland has branched further and lumens within the central ducts are now observed (fig. 3c, d). At this time point the lizard dental gland sits very close to the supralabial gland [26, 28]. From histology the two glands can be distinguished in the adult reptile by differences in morphology and structure (fig. 4a).

In mammals a second lamina is also found developing in close contact with the dental lamina, and, like the dental gland, may form from the same primordium as the dental lamina [29]. This is known as the vestibular lamina and forms the cheek and lip furrow. The vestibular lamina is not continuous around the jaw but is made up of a series of transient bulges and ridges, regionalized in parallel with the dentition. Its initial development is therefore similar to that of the reptilian dental gland, and the two structures have been suggested to share some homology [30]. The vestibular lamina, however, does not undergo branching morphogenesis, but cleaves to create a furrow, known as the oral vestibule. Whether this process involves programmed cell death (apoptosis) is unclear [29, 31]. The function of these additional laminae which form next to the dental lamina are therefore very different.

In most poisonous snakes only one modified dental gland is found on each side of the upper jaw associated with the fangs. Some snake species, however, have a complex of as many as four glands on each side of the head [32]. In keeping with this, multiple dental glands associated with the teeth have been described during embryonic development [33]. These additional glands disappear in most snakes during development leaving only the posteriorly positioned venom gland. The ancestral condition may, therefore, have been for a series of venom glands associated with the maxillary teeth [34]. In non-poisonous snakes dental glands do not appear to form, for example no dental glands are observed in boa constrictors [25].

In tubular-fanged snakes, such as the vipers and elaphids, it is often presumed that the duct of the venom gland extends into the lumen of the tooth. In fact, the duct of the venom glands sits at the base of the tooth and does not extend into the lumen of the gland. When the fang is replaced the new fang moves into position and becomes ankylosed to the maxilla. In this position the duct of the venom gland is in the correct position to allow venom to be channelled into the tooth.

The close association of the dental glands and the teeth in reptiles emphasizes the similarity of these structures during early development. Both teeth and salivary glands develop from epithelial mesenchymal interactions involving the oral epithelium and the neural crest-derived mesenchyme. The early signalling molecules involved in budding and outgrowth of these two organs are also highly conserved, including the Bmps, Fgfs, Wnts and Shh. Loss of Shh for example leads to defects in both tooth and salivary gland development in the mouse [35, 36]. What makes the difference between the teeth and the glands is therefore an interesting question. In the reptiles what signals lead an oral lamina to form teeth, while an adjacent lamina starts to branch and from glandular tissue?

Conclusions

In this chapter some interesting salivary gland adaptations have been discussed. Such adaptations are found throughout the animal kingdom in the vertebrates and invertebrates, similar adaptations having developed independently a number of times. Salivary glands therefore represent very flexible organs, the size, position, and secretions of which can be adapted to create a large range of specialized functions, allowing them to be used for multiple tasks, from making silk to making venom.

Acknowledgement

Thanks to Olda Zahradnicek for pit viper and chameleon samples shown in this chapter.

References

1 Hofmann RR, Streich WJ, Fickel J, Hummel J, Clauss M: Convergent evolution in feeding types: salivary gland mass differences in wild ruminant species. J Morphol 2008;269:240–257.

2 Tandler B, Gresik EW, Nagato T, Phillips CJ: Secretion by striated ducts of mammalian major salivary glands: review from an ultrastructural, functional, and evolutionary perspective. Anat Rec 2001;264:121–145.

3 Ziswiler V, Farner D: Digestion and the Digestive System. New York, Academic Press, 1972.

4 Almansour MI, Jarrar BM: Morphological, histological and histochemical study of the lingual salivary glands of the little egret, *Egretta garzetta*. Saudi J Biol Sci 2007;14:75–81.

5 Parthasarathy R, Gopinathan KP: Comparative analysis of the development of the mandibular salivary glands and the labial silk glands in the mulberry silkworm, *Bombyx mori*. Gene Expr Patterns 2005;5:323–339.

6 Kokubo H, Ueno K, Amanai K, Suzuki Y: Involvement of the *Bombyx scr* gene in development of the embryonic silk gland. Dev Biol 1997;186:46–57.

7 Toyama K: Studies on *Bombyx mori* Eggs. Tokyo, Maruyama-Sah Press, 1909.

8 Bello B, Couble P: Specific expression of a silk-encoding gene of *Bombyx* in the anterior salivary gland of drosophila. Nature 1990;346:480–482.

9 Panzer S, Weigel D, Beckendorf SK: Organogenesis in *Drosophila melanogaster*: Embryonic salivary gland determination is controlled by homeotic and dorsoventral patterning genes. Development 1992; 114:49–57.

10 Gomes G, Silva-Zacarin EC, Zara FJ, Silva de Moraes RL, Caetano FH: Macromolecular array patterns of silk gland secretion in social hymenoptera larvae. Genet Mol Res 2004;3:309–322.

11 Sankaran R: The status and conservation of the edible-nest swiftlet (*Collocalia fuciphaga*) in the Andaman and Nicobar Islands. Biol Conserv 2001; 97:283–294.

12 Marshall AJ, Folley SJ: The origin of nest-cement in edible-nest swiftlets (*Collocalia* spp). Proc Zool Soc London 1956;126:383–389.

13 Medway L: The relationship between the reproductive cycle, moult and changes in the sublingual glands of the swiftlet *Collocalia maxima*. Proc Zool Soc London 1962;138:305–314.

14 Johnston DW: Sex and age characters and salivary glands of the chimney swift. Condor 1958;60:73–84.

15 Ueda A, Nagai H, Ishida M, Nagashima Y, Shiomi K: Purification and molecular cloning of SE-cephalotoxin, a novel proteinaceous toxin from the posterior salivary gland of cuttlefish, *Sepia esculenta*. Toxicon 2008;52:574–581.

16 Yotsu-Yamashita M, Mebs D, Flachsenberger W: Distribution of tetrodotoxin in the body of the blue-ringed octopus *(Hapalochlaena maculosa)*. Toxicon 2007;49:410–412.

17 Gennaro JF Jr, Lorincz AE, Brewster HB: The anterior salivary gland of the octopus *(Octopus vulgaris)* and its mucous secretion. Ann NY Acad Sci 1965; 118:1021–1025.

18 Fry BG, Roelants K, Norman JA: Tentacles of venom: toxic protein convergence in the Kingdom Animalia. J Mol Evol 2009;68:311–321.

19 Kita M, Okumura Y, Ohdachi SD, Oba Y, Yoshikuni M, Nakamura Y, Kido H, Uemura D: Purification and characterisation of blarinasin, a new tissue kallikrein-like protease from the short-tailed shrew *Blarina brevicauda*: comparative studies with blarina toxin. Biol Chem 2005;386:177–182.

20 McDowell SB: The Greater Antillean insectivores. Bull Am Mus Nat Hist 1958;115:115–214.

21 Fox RC, Scott CS: First evidence of a venom delivery apparatus in extinct mammals. Nature 2005;435: 1091–1093.

22 Cuenca-Bescos G, Rofes J: First evidence of poisonous shrews with an envenomation apparatus. Naturwissenschaften 2007;94:113–116.

23 Dufton MJ: Venomous mammals. Pharmacol Ther 1992;53:199–215.

24 Zahradnicek O, Horacek I, Tucker AS: Viperous fangs: development and evolution of the venom canal. Mech Dev 2008;125:786–796.

25 Vonk FJ, Admiraal JF, Jackson K, Reshef R, de Bakker MA, Vanderschoot K, van den Berge I, van Atten M, Burgerhout E, Beck A, Mirtschin PJ, Kochva E, Witte F, Fry BG, Woods AE, Richardson MK: Evolutionary origin and development of snake fangs. Nature 2008;454:630–633.

26 Fry BG, Vidal N, Norman JA, Vonk FJ, Scheib H, Ramjan SF, Kuruppu S, Fung K, Hedges SB, Richardson MK, Hodgson WC, Ignjatovic V, Summerhayes R, Kochva E: Early evolution of the venom system in lizards and snakes. Nature 2006; 439:584–588.

27 Fry BG, Wroe S, Teeuwisse W, van Osch MJ, Moreno K, Ingle J, McHenry C, Ferrara T, Clausen P, Scheib H, Winter KL, Greisman L, Roelants K, van der Weerd L, Clemente CJ, Giannakis E, Hodgson WC, Luz S, Martelli P, Krishnasamy K, Kochva E, Kwok HF, Scanlon D, Karas J, Citron DM, Goldstein EJ, McNaughtan JE, Norman JA: A central role for venom in predation by *Varanus komodoensis* (Komodo dragon) and the extinct giant *Varanus (megalania) priscus*. Proc Natl Acad Sci USA 2009; 106:8969–8974.
28 Kochva E: Oral glands of the reptilia. Biol Reptil 1970;8:43–161.
29 Witter K, Pavlikova H, Matulova P, Misek I: Relationship between vestibular lamina, dental lamina, and the developing oral vestibule in the upper jaw of the field vole (*Microtus agrestis*, Rodentia). J Morphol 2005;265:264–270.
30 Hovorakova M, Lesot H, Peterka M, Peterkova R: The developmental relationship between the deciduous dentition and the oral vestibule in human embryos. Anat Embryol (Berl) 2005;209:303–313.
31 Kim JG, Iwailiao Y, Higashi Y: Observations on early development of the murine fetal oral vestibule. Okajimas Folia Anat Jpn 1998;75:131–139.
32 Kochva E, Gans C: Salivary glands of snakes. Clin Toxicol 1970;3:363–387.
33 Martin H: Etude de l'appareil glandulaire venimeux chez un embryon de *Vipera aspis*. Bull Soc Zool Fr 1899;24:106–116.
34 Phisalix M: Animaux venimeux et venins. Paris, Masson, 1922, vol 2.
35 Jaskoll T, Leo T, Witcher D, Ormestad M, Astorga J, Bringas P Jr, Carlsson P, Melnick M: Sonic hedgehog signaling plays an essential role during embryonic salivary gland epithelial branching morphogenesis. Dev Dyn 2004;229:722–732.
36 Dassule HR, Lewis P, Bei M, Maas R, McMahon AP: Sonic hedgehog regulates growth and morphogenesis of the tooth. Development 2000;127:4775–4785.

Abigail S. Tucker, Senior Lecturer
Department of Craniofacial Development and Orthodontics
Floor 27, Guy's Hospital, London SE1 9RT (UK)
Tel. +44 207 188 8038, Fax +44 207 188 1674
E-Mail Abigail.tucker@kcl.ac.uk

Tucker AS, Miletich I (eds): Salivary Glands. Development, Adaptations and Disease.
Front Oral Biol. Basel, Karger, 2010, vol 14, pp 32–47

Genetic Regulation of Salivary Gland Development in *Drosophila melanogaster*

Carolyn Pirraglia · Monn Monn Myat

Weill Medical College of Cornell University, New York, N.Y., USA

Abstract

Studies of salivary gland development in the *Drosophila* embryo have revealed the morphogenic events by which the salivary gland tubes are formed, and identified the genes and genetic networks that regulate these events. Invagination of the salivary gland primordium occurs by an apical constriction mechanism regulated by the tyrosine kinase, Tec29, the transcription factors, Fork head and Huckebein, and Rho GTPase-mediated actomyosin contraction. After invagination is complete, transcriptional control of the apical membrane protein, Crumbs, by the transcription factors, Hairy, Hkb and Ribbon, and downregulation of Moesin-dependent apical stiffness promotes elongation of the salivary gland lumen. Integrin-mediated adhesion between the gland and surrounding mesoderm, coupled with Rho GTPase-mediated contraction of the proximal gland cells and downregulation of E-cadherin-based cell-cell adhesion by the Rac GTPases, allow turning and posterior migration of the salivary gland. Further posterior migration of the salivary gland is dependent on axon guidance cues, such as Robo and Slit, and close association with surrounding tissues. Many of the genes identified as regulators of salivary gland invagination and migration also control invagination of other epithelial tissues and migration of diverse cell types. Thus, studies of *Drosophila* salivary gland morphogenesis continue to illuminate the conserved mechanisms by which cells give rise to three-dimensional tissues and organs during embryogenesis.

During embryonic development, cells acquire distinct fates and undergo complex morphogenic movements to form three-dimensional tissues and organs. In recent years, much attention has been placed on understanding the cellular and molecular mechanisms of tubular organ formation during development of both vertebrate and invertebrate organisms. The developing salivary gland of the fruit fly, *Drosophila melanogaster*, is a widely-used model system for tubulogenesis. Studies on *Drosophila* salivary gland development have yielded valuable insights into how cell fates are distinguished within an organ primordium, how cells change morphology and migrate to form a tubular organ and how tube size and shape are controlled. Many features of *Drosophila* salivary gland morphogenesis are conserved during development of

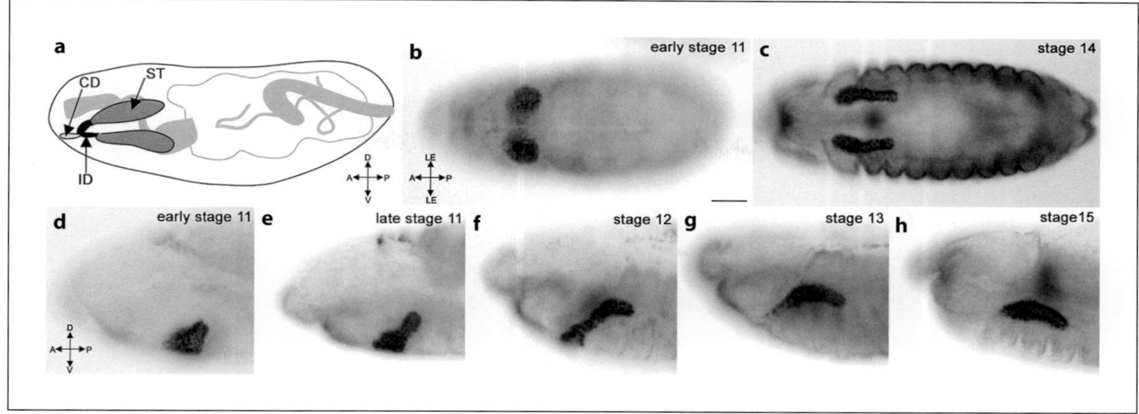

Fig. 1. *Drosophila* embryonic salivary gland morphogenesis. **a** The embryonic salivary gland consists of a pair of elongated secretory tubes (ST) that connect to the larval mouth via the finer individual duct (ID) and common duct (CD) tubes. Salivary gland primordial cells originate as two placodes on the ventral surface of the embryo that invaginate and migrate collectively to form a pair of elongated tubes (**b–h**). **b–h** All embryos were stained for the transcription factor dCreb-A to label the salivary gland nuclei. Embryo orientation is designated by A = anterior, P = posterior, D = dorsal, V = ventral and LE = lateral epidermis. **b** and **c** are ventral views and **d–h** are lateral views. Scale bar represents 20 μm.

other tubular organs, such as the mammalian submandibular gland, kidney, lung and vasculature. Thus, insights gained from *Drosophila* salivary gland development have benefited our general understanding of tubulogenesis.

The *Drosophila* salivary glands are one of the simplest types of tubular organs found in nature. They are involved in secretion and consist of a pair of elongated tubes that are connected to the larval mouth through the individual duct tubes and the common duct tube (fig. 1). Studies of *Drosophila* salivary gland secretion have been limited to glue protein synthesis and secretion in response to the 20E steroid hormone prior to puparium formation, the developmental stage during which glue protein is used to attach the cocoon to a solid support [1]. As expected for a secretory organ, expression of genes encoding proteins required for the early steps of the secretory pathway, such as the targeting and translocation of proteins to the ER, and the transport of proteins between the ER and Golgi, are elevated in the early embryonic salivary gland [2].

The larval salivary glands are formed during mid-embryogenesis as two clusters of approximately 100 cells each on the ventral surface of the embryo within parasegment two (fig. 1). Specification of salivary gland cell fate requires the homeotic genes *sex combs reduced (scr)*, *extradenticle (exd)* and *homothorax (hth)*, as well as signaling by the TGF-β homolog, encoded by *decapentaplegic (dpp)* [3–5]. In embryos mutant for Scr, Exd or Hth, no salivary glands are formed [3–5]. Within each salivary gland placode, the ventral-most cells closest to the ventral midline become specified as salivary duct cells (both common and individual ducts) in a process dependent on EGFR signaling, whereas the remaining cells within the salivary gland placode adopt a secretory cell fate [6, 7]. Once the salivary gland secretory and duct fates are established, there is no further cell division or cell death, and all subsequent morphogenic events that drive tubulogenesis are dependent upon the cells that comprise the salivary gland primordium. Salivary gland morphogenesis begins with the invagination of the secretory primordial cells to form a pair of elongated tubes, each of which consists of a monolayer of cells surrounding a central lumen (fig. 1). Subsequent to invagination, the internalized secretory tubes migrate towards the posterior of the embryo as an intact organ until they reach their final position along the lateral body wall. Concomitant with posterior migration of the secretory tubes, the salivary duct cells invaginate to form a connected network of tubes by the end of embryogenesis.

The amenability of the *Drosophila* embryo to cell biological techniques, such as immunocytochemistry, histological sectioning and imaging by confocal microscopy, have allowed a detailed analysis of the morphogenic movements by which salivary gland primordia form the secretory and duct tubes. Studies of *Drosophila* salivary gland morphogenesis also benefit from the wide range of available genetic techniques that allow identification of the genes and genetic networks required for salivary gland development. For example, P-element insertion screens and chemical mutagenesis screens have identified genes expressed in the salivary glands and genes required for salivary gland development. In addition, the UAS-GAL4 system [8] readily allows the investigator to test whether a gene of interest is required cell-autonomously, such as within the salivary gland cells, or non-cell-autonomously, such as within the surrounding tissues, for gland development. Lastly, the advent of modern imaging techniques, such as multi-photon microscopy, allows live visualization of salivary gland migration as it occurs in a developing embryo. Here, we review our most recent understanding of *Drosophila* embryonic salivary gland development, in particular, the morphogenic movements by which the secretory gland cells invaginate to form a tubular organ and how the gland migrates cohesively as an intact organ. For a complete list of genes required for *Drosophila* salivary gland development, see table 1.

Salivary Gland Invagination

Secretory cells of the embryonic salivary gland undergo orchestrated cell shape changes to invaginate and form the elongated secretory tubes. During gland invagination, the columnar epithelial cells of the secretory gland (hereafter referred to as the salivary gland) undergo morphological changes to become wedge-shaped in a process driven by the constriction of the apical domain and basal migration of nuclei. Constriction of the apical membrane during salivary gland invagination occurs by a similar mechanism to that of other invaginating epithelia, such as the vertebrate neurectoderm [9] and the *Drosophila* ventral epidermis [10, 11]. Fork head (Fkh), a

Table 1. Genes required for salivary gland morphogenesis

Gene	Salivary gland morphogenesis	Specific cellular event
tec29	invagination	endocycle regulation/ actin organization
chickadee (chic)	invagination	actin organization
twinstar (tsr)	invagination	actin organization
fork head (fkh)	invagination	apical constriction/cell survival
faint sausage (fas)	invagination	order of SG cell invagination
huckebein (hkb)	invagination	order of SG cell invagination/ apical membrane regulation
hairy	invagination	apical membrane regulation
klarsicht (klar)	invagination	apical membrane regulation
crumbs (crb)	invagination	apical membrane regulation
18 wheeler (18w)	invagination	apical constriction
rho1 kinase (rok)	invagination	apical constriction/apical-basal contraction of proximal cells
rho1 GTPase	invagination/ posterior migration	apical constriction/apical polarity/apical-basal contraction
twist (twi)	invagination/ posterior migration	mesoderm development
daughterless (da)	invagination/ posterior migration	mesoderm development
ribbon (rib)	posterior migration	apical membrane regulation
heartless (htl)	posterior migration	visceral mesoderm development
multiple edematous wings (mew)	posterior migration	integrin-mediated adhesion
inflated (if)	posterior migration	integrin-mediated adhesion
myospheroid (mys)	posterior migration	integrin-mediated adhesion
rac GTPases (rac1, rac2 and mtl)	posterior migration	regulation of E-cadherin mediated cell-cell adhesion
β_2-tubulin (β2t)	posterior migration	Fusion competent myoblast (FCM) and Longitudinal visceral mesoderm founder cell (LVMF) migration/ integrin-mediated adhesion
jellybelly (jeb)	posterior migration	LVMF migration
binous (bin)	posterior migration	LVMF migration

Table 1. Continued

Gene	Salivary gland morphogenesis	Specific cellular event
netrin A (net) and B/frazzled (fra)	posterior migration	guidance chemoattractant/receptor
slit/roundabout (robo) 1, 2 and 3	posterior migration	guidance chemorepellent/receptor
wnt5/derailed (drl)	posterior migration	guidance chemorepellent/receptor
wnt4/frizzled (fz) 1 and 2	posterior migration	guidance chemorepellent/receptor
pvf1 and pvf2/pvr	posterior migration	guidance chemorepellent/receptor
serpent (srp)	posterior migration	fat body development
gooseberry (gsb)	posterior migration	fat body development
eyes absent (eya)	posterior migration	fat body development

transcription factor that contains a winged-helix DNA-binding domain, is required for the apical membrane constriction of gland cells. In the absence of Fkh function, gland cells fail to constrict their apical domains resulting in the failure of the salivary gland cells to internalize [12]. In addition, Fkh is required for the survival of salivary gland cells by preventing apoptosis [12], which is in part mediated by the zinc-finger transcription factor, Senseless (Sens) [13].

Salivary gland cells invaginate in a regulated and sequential manner, beginning with the dorsal-posterior cells, then the dorsal-anterior, followed by the ventral-anterior and, lastly, the ventral-posterior cells of the placode. The sequential invagination of gland cells plays an important role in determining the final shape and size of the mature gland. *huckebein (hkb)*, which encodes an Sp1/Egr-like transcription factor, controls the sequence of invagination in the salivary gland placode [14]. In *hkb* mutant embryos, the sequence of internalization is aberrant, resulting in the formation of dome-shaped glands instead of the elongated glands observed in wild-type [14]. A putative cell adhesion protein, encoded by the *faint sausage (fas)* gene, also controls the order of salivary gland cell invagination; in *fas* mutant embryos, dome-shaped glands, like those of *hkb* mutant embryos, are formed [14].

By the time the salivary gland placode is formed, gland cells have already established apical-basal (Ap-Bl) polarity and this polarity is maintained throughout gland morphogenesis. Failure to maintain Ap-Bl polarity abrogates gland morphogenesis as observed in embryos mutant for the small GTPase Rho1. Rho1 regulates polarity of the salivary gland epithelium by specifically maintaining apical localization of the membrane proteins, Crumbs (Crb), Stardust (Sdt) and atypical protein kinase (aPKC). Rho1 performs this function, in part, through its regulation of the transcription and apical localization of *crb* mRNA [15]. In addition to maintaining Ap-Bl polarity, Rho1 controls apical constriction and cell shape changes of gland cells through Rho1 kinase (Rok) and actomyosin contraction [15]. Moreover, the Toll-like receptor, 18 wheeler (18W), regulates gland invagination by positively regulating Rho GTPase activity, possibly through inactivation of the negative regulators of the Rho GTPase signaling pathway, such as RhoGAPs [16].

Actin reorganization and actomyosin contraction in the salivary gland cells is important for proper gland development. In particular, accumulation of Filamentous (F)-actin at the apical surface of invaginating salivary gland cells is important for gland internalization and has been shown to be dependent on the non-receptor tyrosine kinase, Tec29. In *tec29* mutant embryos, there is a reduction in apical F-actin and an increase in monomeric Globular (G)-actin, resulting in a delay in salivary gland invagination [17]. The invagination defect observed in *tec29* mutants is enhanced by a mutation in Chickadee (Chic), the *Drosophila* homolog of Profilin, which is required for actin polymerization and F-actin formation [17]. In contrast, a mutation in Twinstar (Tsr), the *Drosophila* homolog of Cofilin, which increases G-actin by promoting actin depolymerization, suppresses the *tec29* mutant invagination defect [17]. Therefore, Tec29 is required for the formation and/or maintenance of F-actin at the apical membrane of gland cells during invagination.

Concomitant with salivary gland invagination, gland cells enter a specialized cell cycle, the endocycle (endoreplication), in which they cycle between G and S phase without cytokinesis leading to polyploidy [18]. Immediately following their invagination, the salivary gland cells enter the endocycle, which proceeds as a wave from the distal-tip to the proximal-end of the gland [17, 18]. Tec29, in addition to promoting apical F-actin accumulation as described above, also plays an important role in the regulation of the endocycle in salivary gland cells. In *tec29* mutants, the synchronized wave of endocycle progression is lost, resulting in uncoordinated and premature endoreplication throughout the gland [17]. Blocking the ability of gland cells to enter the endocycle by continuous expression of Cyclin E, an S-phase regulator, in the gland primordium, suppresses the *tec29* mutant invagination phenotype [17]. These data indicate that coordinated endoreplication is necessary for salivary gland invagination and that Tec29 regulates the progression of this endocycle wave throughout the gland primordium. In addition, a mutation in Chic enhances the endoreplication defect of *tec29* mutants, suggesting that remodeling of the actin cytoskeleton plays a role in endocycle regulation in the salivary gland [17].

Salivary Gland Lumen Elongation

Studies in the *Drosophila* salivary gland were the first to show that genesis of apical membrane is a key mechanism by which the gland lumen elongates during gland morphogenesis. Three transcription factors, Hairy, Hkb and Ribbon (Rib), play key roles in controlling lumen elongation during gland development. Hairy, a basic helix-loop-helix (bHLH) transcription factor, has been shown to regulate apical membrane growth, partly through its regulation of Hkb [19]. In *hairy* mutant embryos, the normally unbranched and elongated gland lumena are expanded and/or branched, and *hkb* RNA expression fails to be properly temporally repressed [19]. In *hkb* mutant embryos, salivary gland cells have reduced apical membranes and the salivary gland lumen fails to elongate, whereas overexpression of Hkb, specifically in the gland, increases apical membrane length and expands the gland lumen [19]. Hkb promotes apical membrane growth through transcriptional upregulation of two downstream target genes, *klarsicht (klar)*, which encodes a regulator of microtubule minus-end transport [20], and *crb,* which encodes an apical membrane protein that confers apical membrane identity [21–23]. Upregulation of Crb in salivary gland cells promotes apical membrane growth, whereas Klar functions to mediate delivery of vesicles to the apical membrane, thereby allowing elongation of the gland lumen. These data indicate that Hairy limits apical membrane growth during invagination through transcriptional repression of *hkb*, an important regulator in apical membrane organization [19].

In addition to Hkb, *crb* transcription in salivary gland cells is also dependent on Rib, a broad tramtrack bric-a-brac (BTB)-domain containing nuclear protein that functions with its binding partner, Lola-like (Lolal), also a BTB-domain containing protein [24]. Rib controls salivary gland lumen elongation by increasing *crb* transcript levels and downregulating active, phosphorylated Moesin (Moe), a member of the Ezrin-Radixin-Moesin (ERM) family of proteins that localizes predominantly to the apical surface and functions to link the actin cytoskeleton to the apical membrane [24]. Thus, in *rib* mutant glands, transcriptional levels of *crb* are reduced and Moe activity at the apical membrane is increased compared to wild-type [24]. Deep-tissue live imaging of wild-type and *rib* mutant glands using multi-photon microscopy revealed slow and incomplete lumen growth in *rib* mutant glands [25], suggesting that apical domains of *rib* mutant salivary glands are rigid and more resistant to morphological changes than those of wild-type glands. This hypothesis was further explored by generating computational models of lumen elongation based on live imaging, which suggest that *rib* mutant glands show a three- to fivefold increase in apical stiffness and a twofold increase in apical viscosity compared to wild-type, resulting in a slower rate of luminal elongation [25]. These studies in the *Drosophila* salivary gland collectively show that lumen and tube elongation are dependent on apical membrane growth and delivery, and the interactions between the apical membrane and cytoskeleton that affect apical stiffness and viscosity.

Salivary Gland Migration

During embryogenesis, many cells migrate cohesively or collectively as an intact tissue. Although much is known about how single cells migrate from cell culture

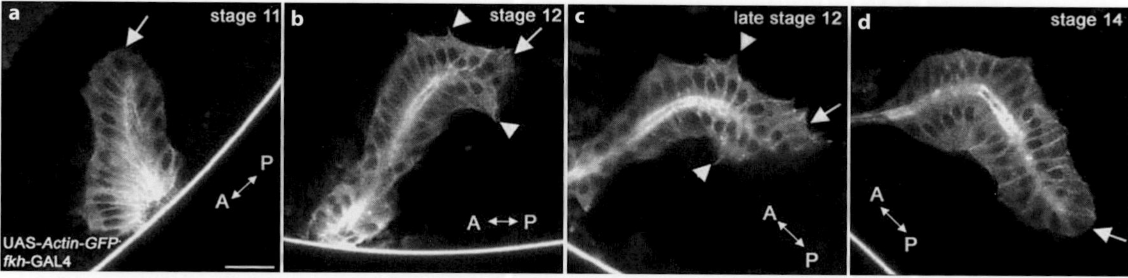

Fig. 2. Collective migration of the *Drosophila* embryonic salivary gland. In wild-type embryos, prior to contact with the circular visceral mesoderm (CVM), the salivary gland cells (**a**) do not protrude membrane extensions (**a**, arrow), whereas upon migration along the CVM (**b**), the distal gland cells form lamellipodia (**b**, arrow) and filopodia (**b**, arrowheads). During late stage 12 (**c**), as more cells contact the CVM and migrate, more filopodia (**c**, arrowheads) and lamellipodia (**c**, arrow) are observed. During stage 14 (**d**), when most of the gland has turned and migrated posteriorly, gland cells no longer protrude filopodia or lamellipodia (**d**, arrow). The orientation of the embryo is designated in each panel by A = anterior and P = posterior. Scale bar represents 10 μm.

studies, it remains elusive how epithelial cells, with their prominent zones of cell-cell adhesion, migrate as a cohesive group or sheet. Studies in the *Drosophila* embryonic salivary gland have contributed significantly to our understanding of collective cell migration. Live imaging studies revealed that cells in the distal half of the salivary gland migrate by extending membrane protrusions and elongating in the direction of migration, whereas cells in the proximal half of the gland migrate by contracting along the Ap-Bl axis and changing shape from columnar to cuboidal (fig. 2). In addition to maintaining Ap-Bl polarity during salivary gland invagination as mentioned above, Rho1 GTPase is required in the proximal cells of the salivary gland to promote Ap-Bl contraction and regulate the change in cell shape from columnar to cuboidal during gland migration. Rho1 mediates these effects, at least in part, through activation of Rok and the actomyosin contractile machinery [15].

An unanswered question, regarding not only salivary gland migration but collective migration in general, is how epithelial cells migrate while remaining attached to neighboring cells via cell-cell junctions. Throughout invagination, turning and posterior migration, salivary gland cells are adhered to one another through E-cadherin (E-cad)-based adherens junctions. The small GTPase, Rac, has been shown to be critical in regulating E-cad levels during salivary gland migration. In embryos mutant for the three *Drosophila rac* genes, *rac1*, *rac2* or *mtl*, E-cad-mediated adhesion is increased and glands fail to complete invagination and migrate posteriorly, whereas activation of Rac1 in gland cells downregulates E-cad-mediated adhesion and leads to dispersal of gland cells [26]. One mechanism by which the Rac GTPases downregulate E-cad in salivary gland cells is through dynamin-mediated endocytosis [26], although other mechanisms of E-cad regulation have not been ruled out. These results indicate that

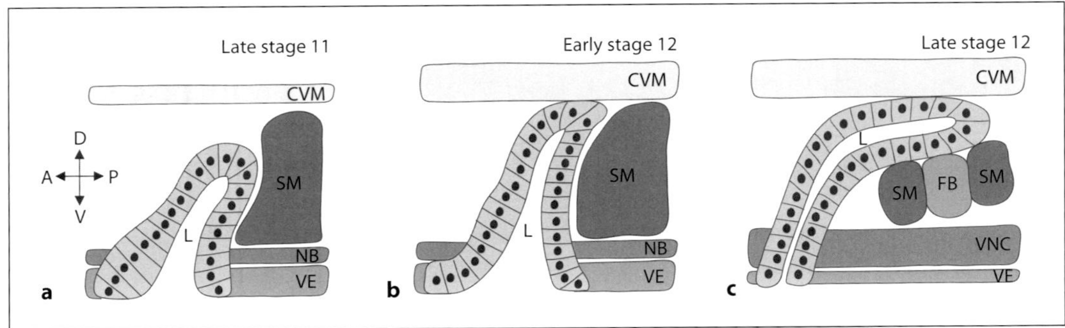

Fig. 3. Migration of the *Drosophila* embryonic salivary gland along surrounding tissues. During invagination from the ventral epidermis (VE), wild-type salivary glands migrate dorsally through clusters of neuroblast (NB) cells and somatic mesoderm (SM) to contact the overlying circular visceral mesoderm (CVM) with their distal-most cells (**a**, **b**). During late stage 12, the salivary glands migrate posteriorly over clusters of SM and fat body (FB) cells to achieve proper positioning (**c**). The orientation of the embryo is designated by A = anterior, P = posterior, D = dorsal and V = ventral. L = lumen. NBs begin to differentiate and become part of the developing ventral nerve cord (VNC) during late stage 12.

dynamic regulation of E-cad-mediated adherens junctions is essential for proper salivary gland development.

Like most migrating tissues, the salivary gland migrates by interacting intimately with surrounding mesoderm. Throughout invagination and migration, the salivary gland directly contacts five distinct embryonic tissues: the circular visceral mesoderm (CVM), the VM-derived gastric caecae (GC), the somatic mesoderm (SM), the fat body (FB) and the neurectoderm-derived neuroblasts (NBs). Mutations that affect the development of these tissues result in defects in salivary gland positioning. Invaginating gland cells first migrate through the NBs of the developing ventral nerve cord (VNC) before interacting with SM precursor cells (fig. 3). Gland cells then migrate through clusters of SM precursor cells to contact the CVM to initiate posterior migration (fig. 3). As the glands migrate posteriorly, the ventral surface of the salivary glands remain associated with clusters of SM cells (fig. 3). Mesodermal cell fate is regulated by the bHLH domain-containing protein, Twist (Twi), which exists as homodimers or heterodimers with its binding partner, Daughterless (Da) [27]. The resulting dimers bind DNA and function as positive or negative regulators of transcription. The ectopic expression of Twi-Twi homodimers throughout the SM induces an increase in the number of SM cells resulting in significantly overelongated glands [28]. Conversely, a reduction in the number of SM cells, by expression of Twi-Da heterodimers throughout the SM, results in salivary gland invagination and migration defects, indicating that proper SM formation is required for gland development [28]. In addition to interacting with clusters of SM cells, the ventral side of the migrating salivary gland also contacts the FB, a mesodermally-derived tissue that exists as bilateral cell clusters located between SM cell clusters (fig. 3). Mutations in genes required for FB development, such as *serpent (srp)*, *gooseberry (gsb)* and *eyes absent (eya)*, result in mispositioning of the salivary gland with the distal-tip oriented towards the ventral epidermis, instead of lying parallel along the anterior-posterior (A-P) axis as in wild-type [28].

Cells at the distal-tip of the salivary gland are the first to contact the overlying CVM after invagination (fig. 3). The close association of the distal gland cells and the CVM is dependent on integrin-mediated adhesion. Integrins function as heterodimers of two transmembrane proteins: an α and a β subunit. In *Drosophila*, there are 5 α subunits (αPS1–5) and 2 βPS subunits (βPS and βv). During gland development, the salivary gland expresses αPS1βPS, whereas the surrounding mesoderm expresses αPS2βPS [29]. In embryos mutant for either αPS1, encoded by *multiple edematous wings (mew)*, or αPS2, encoded by *inflated (if)*, salivary gland cells invaginate properly; however, glands fail to contact the CVM and migrate posteriorly resulting in curved or folded glands [29]. In addition, maternal and zygotic loss of βPS, which is encoded by *myospheroid (mys)*, results in salivary gland defects similar to those observed in *αPS2* and *αPS1* zygotic mutant embryos [29]. Although it is clear that integrin function is critical for establishing contact between the gland and CVM, little is known about how integrin expression and function are controlled during gland migration. Interestingly, mutations in the beta-2 tubulin isoform (β2t), a subunit of microtubules that are involved in intracellular transport, disrupts integrin accumulation at contact sites between the gland and CVM, and the gland and the surrounding SM, thus impeding posterior migration of the gland [30]. These data suggest that one likely mechanism of integrin regulation is the transport of integrin subunits to sites of adhesion during gland migration.

In addition to providing integrin-mediated adhesion to promote posterior migration, the CVM functions as a physical barrier for proper gland positioning. Mutations in genes important for CVM formation and organization indirectly affect salivary gland migration. For example, *heartless (htl)*, which encodes an FGF receptor required for the migration and differentiation of mesodermal cells, is required for normal salivary gland migration. In *htl* mutants, the CVM is discontinuous, existing as clusters of tissue as compared to the continuous CVM structure observed in wild-type; therefore, in *htl* mutants, the gland migrates dorsally between the CVM clusters instead of turning and migrating posteriorly [29].

After the salivary gland turns posteriorly, it detaches from the CVM and continues migrating along the A-P axis. Gland detachment from the CVM occurs concomitantly with the ingression of surrounding tissues between the gland and the CVM. For example, fusion competent myoblasts (FCMs) and longitudinal visceral mesoderm founder cells (LVMFs) migrate between the CVM and the distal half of the gland that contacts the CVM. A number of genes that have been identified to control either FCM or LVMF migration have also been shown to indirectly regulate gland migration. In embryos mutant for *jellybelly* (*jeb*) or *binous* (*bin*), genes that play a role

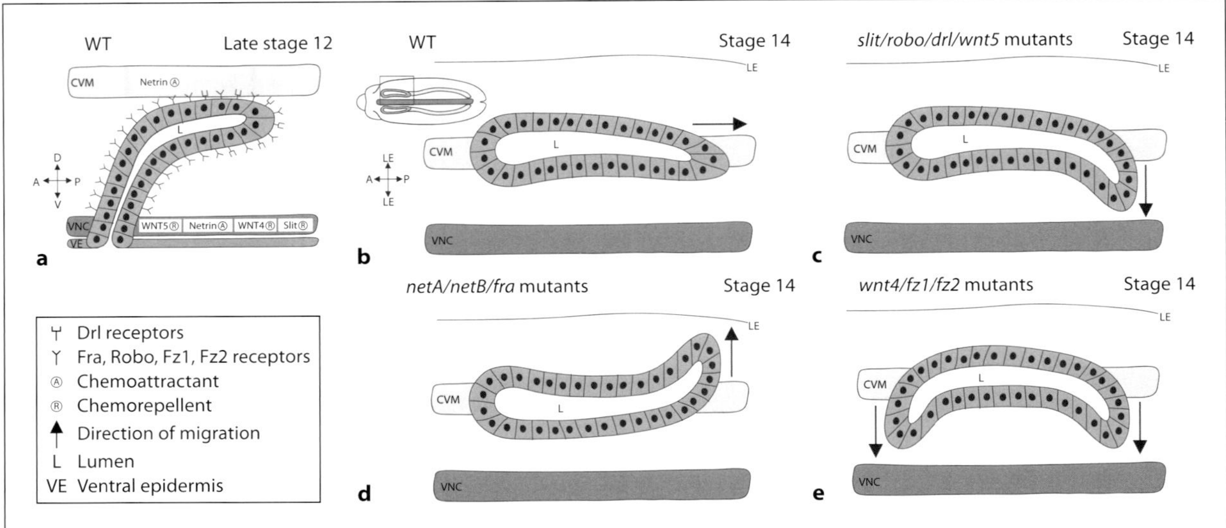

Fig. 4. Role of axonal guidance cues during *Drosophila* embryonic salivary gland migration. In wild-type embryos (**a**) during late stage 12, the chemoattractant Netrin is expressed in the circular visceral mesoderm (CVM) and the developing ventral nerve cord (VNC), which also expresses the chemorepellants, Slit, Wnt4 and Wnt5, to promote proper orientation of salivary glands at stage 14 (**b**). In embryos mutant for Slit, Robo, Drl or Wnt5 (**c**), the distal-tip cells of the salivary glands migrate towards the VNC. In *netA*, *netB* or *fra* mutant embryos (**d**), the distal-tip cells of the salivary gland are deflected away from the VNC and CVM towards the lateral epidermis (LE). In embryos mutant for Wnt4, Fz1 or Fz2 (**e**), the proximal and distal cells of the salivary glands migrate towards the VNC. The orientation of the glands in A-E are designated by A = anterior, P = posterior, D = dorsal, V = ventral and LE = lateral epidermis. Glands in **b–e** are enlarged images of the selected region of the embryo in **b**.

in LVMF migration, the gland does not detach from the CVM and becomes overextended as the CVM retracts posteriorly to become part of the gut [28]. Similarly, in *β2t* mutant embryos, FCMs and LVMFs do not migrate between the CVM and salivary gland, delaying detachment of the gland from the CVM and delaying further posterior migration [30].

Axon Guidance Cues in Salivary Gland Migration

As described above, posterior turning of the salivary gland at the contact point between the distal-tip cells and the overlying CVM is dependent on proximal cell contraction, integrin-mediated adhesion and lumen elongation. Beyond the contact point, several guidance cues have been identified to control further posterior migration of the gland (fig. 4). Interestingly, these guidance cues were originally identified in axonal migration, suggesting that salivary gland collective migration and axonal migration share common molecular mechanisms. Two previously identified axon guidance cues, Netrin and Slit, have been shown to guide salivary gland migration (fig. 4). In *Drosophila* there are two functionally redundant *netrin* genes, *netA* and *netB*, which encode secreted guidance molecules. During gland migration, NetA and NetB are expressed in the VNC and the CVM (fig. 4). Ectopic expression of NetB throughout the entire developing VNC using the UAS-GAL4 system results in the misdirection of the salivary gland towards the exogenous Netrin source, whereas loss of NetA and/or NetB function results in the deflection of the salivary gland towards the lateral body wall. These results indicate that Netrins act as attractive guidance cues for proper gland positioning [31]. Netrins signal through membrane receptors, of which two are expressed in *Drosophila*, Frazzled (Fra) and Unc-5. Fra is expressed in invaginating salivary gland cells and mutations in Fra result in gland migration defects similar to those observed in *netA* and *netB* mutants [31]. The lateral deflection of the salivary gland in the *netrin* mutants is due, in part, to the activity of the chemorepellent Slit, which is secreted by the VNC (fig. 4) [31–33]. In *slit* mutant embryos, the glands curve medially, towards the ventral midline, whereas misexpression of Slit in the CVM results in the glands migrating ventrally away from the CVM [31]. Slit signaling is mediated by the Roundabout (Robo) receptors, which are encoded by *robo1*, *robo2* and *robo3*. Robo1 and Robo2 are endogenously expressed in the salivary gland and when mutated cause gland migration defects similar to those of *slit* mutants [31].

Canonical Wnt signaling, which is meditated by the Frizzled (Fz) receptors, is involved in many fundamental processes ranging from cell fate determination to axon guidance [34, 35]. During salivary gland development, Wnt4/Fz signaling has been shown to be required for proper gland migration. Mutations in Wnt4, which is expressed in the VNC and in narrow ectodermal stripes, or mutations in the Fz receptors, Fz1 or Fz2, which are expressed in the salivary gland, result in misguided glands that aberrantly migrate towards the ventral midline (fig. 4) [36]. In addition to Fz receptors, related-to-tyrosine-kinase (RYK) receptors, such as the atypical Wnt RYK receptor Derailed (Drl), can also mediate Wnt signaling [37]. Like Wnt4 and Fz, Wnt5 and its cognate receptor, Drl, are expressed in the VNC and salivary gland, respectively (fig. 4) [36]. Interestingly, Drl expression is limited to the distal most gland cells and mutations in Wnt5 or Drl result in gland migration defects with the distal-tip cells specifically curving towards the ventral midline [36]. Furthermore, ectopic expression of Wnt4 or Wnt5 in the CVM results in glands migrating away from the ectopic source, indicating that Wnt4 and Wnt5 act as chemorepellents during salivary gland migration. In summary, gland migration requires two Wnt signaling pathways: one pathway mediated by Wnt4/Fz1 and Fz2, which is required throughout the entire gland, and the Wnt5/Drl pathway, which is required specifically within the distal-tip cells of the migrating salivary gland.

In addition to axon guidance cues, the platelet derived growth factor (PDGF) pathway has been shown to regulate gland migration. Pvr, the only known *Drosophila* homolog of the mammalian PDGF receptor, has been previously shown to be important for cell migration of various developing tissues in *Drosophila*, such as the border cells [38]. Pvr and one of the three *Drosophila* PDGF ligands, Pvf1, are expressed in the developing salivary glands, whereas the PDGF ligands, Pvf2 and Pvf3, are expressed by the VNC [39]. Embryos mutant for Pvf1, Pvf2 or Pvr, have glands that curve towards the ventral midline, instead of lying parallel to the A-P axis as in wild-type, indicating that the Pvr/Pvf1 and Pvf2 signaling pathway functions to direct gland migration along the CVM [39].

Salivary Gland Mutagenesis Screens

Large-scale mutagenesis screens have made it possible to identify genes and genetic networks controlling *Drosophila* salivary gland development with relative ease. Studies of *Drosophila* salivary gland development have benefited greatly both from loss-of-function and gain-of-function mutagenesis screens. Two types of loss-of-function mutagenesis screens have been reported: a chemical EMS screen for genes that when mutated disrupt salivary gland and tracheal development [40] and a deficiency screen for genomic intervals that when deleted abrogate salivary gland development [30]. In addition, one gain-of-function screen has been recently performed, which identified numerous genes that when overexpressed result in salivary gland defects.

Due to the ability to track *Drosophila* chromosomes with markers visible in the adult, the EMS mutagenesis screen was designed to only recover those mutations on the third chromosome, the largest of the four chromosomes in *Drosophila*. Although the EMS mutagenesis screen was not carried out to saturation, it generated many mutations that led to the identification of genes not previously known to be required for salivary gland development. For example, the EMS screen identified a novel role for the pair-rule gene, *hairy*, in embryonic salivary gland development as mentioned above [19] and demonstrated a novel function of the RNA-binding protein, Pasilla (Ps), in salivary gland secretion [41]. In addition, the EMS screen revealed a novel requirement for one of the FGF receptors encoded by the *heartbroken (hbr)* gene in salivary gland migration [29] and yielded several new and, as of yet, uncharacterized mutations that affect gland migration.

The deficiency screen was carried out by analyzing 100 *Drosophila* lines, each of which had a defined interval of the third chromosome deleted, and collectively, deleted approximately 92–95% of the third chromosome polytene bands [http://flybase.bio.indiana.edu]. Of the 100 lines, 34 were identified that, when homozygous, showed defects in salivary gland migration. These 34 lines were then further grouped into three classes: (1) where the salivary gland distal-tip did not initiate turning and migration, (2) where only the distal-tip of the gland turned or (3) where the distal half of the gland turned but the proximal half did not [30]. Further analysis of smaller deficiencies that delete genomic intervals contained within one of the 34 genomic intervals led to the identification of a novel role for β2t in salivary gland migration as described above.

A recently reported gain-of-function screen for genes affecting salivary gland development utilized a salivary gland-specific GAL4 driver, *fkh*-GAL4, to drive expression of transposable element (EP element) lines [42]. In such an approach, binding of GAL4 to UAS sites within the EP element, in most cases, leads to overexpression of the downstream gene. This gain-of-function screen identified 44 EP insertions with salivary gland defects, of which 14 corresponded to genes with no previously reported role in salivary gland development. Some of the candidate genes identified are endogenously expressed in the salivary gland, such as *bitesize (btsz)*, which encodes the only *Drosophila* synaptotagmin-like protein. Other genes that were identified are genes that are not endogenously expressed in the gland, such as *capricious (caps)*, which encodes a leucine-rich repeat transmembrane protein [42]. The genes and genomic intervals identified in the screens mentioned above will no doubt lead to a more thorough understanding of the genetic networks that control salivary gland development.

Conclusion

The *Drosophila* embryonic salivary gland has emerged as one of the best model systems to study tubulogenesis. The plethora of available genetic techniques combined with state-of-the-art imaging techniques have led to the identification of conserved genes and cellular mechanisms by which epithelia form tubes. These mechanisms include invagination, lumen elongation and migration. Despite the above-stated advances, much still needs to be learned about salivary gland development and tube formation in general. For example, it is still not clear how adhesion between migrating salivary gland cells is regulated to maintain cohesion of the tube but allow enough plasticity for migration. It is also not clear how the proximal and distal gland cells, with their unique modes of migration, contribute to overall gland migration. Further exploration of the genetic networks already known to control salivary gland development, characterization of genes identified in the mutagenesis screens and computational modeling of gland morphogenesis will undoubtedly bring novel insights in the future.

References

1 Lehmann M: *Drosophila sgs* genes: stage and tissue specificity of hormone responsiveness. Bioessays 1996;1:47–54.

2 Abrams E, Andrew D: CrebA regulates secretory activity in the *Drosophila* salivary gland and epidermis. Development 2005;132:2743–2758.

3 Henderson KD, Andrew DJ: Regulation and function of *scr, exd* and *hth* in the *Drosophila* salivary gland. Dev Biol 2000;217:362–374.

4 Henderson K, Isaac D, Andrew D: Cell fate specification in the *Drosophila* salivary gland: the integration of homeotic gene function with the Dpp signaling cascade. Dev Biol 1999;205:10–21.

5 Panzer S, Weigel D, Beckendorf SK: Organogenesis in *Drosophila melanogaster*: embryonic salivary gland determination is controlled by homeotic and dorsoventral patterning genes. Development 1992; 114:49–57.

6 Kuo YM, Jones N, Zhuo B, Panzer S, Larson V, Beckendorf S: Salivary duct determination in *Drosophila*: roles of the EGF receptor signaling pathway and the transcription factors Fork head and Trachealess. Development 1996;122:1909–1917.

7 Haberman AS, Isaac DD, Andrew DJ: Specification of cell fates within the salivary gland primordium. Dev Biol 2003;258:443–453.

8 Brand AH, Perrimon N: Targeted gene expression as a means of altering cell fates and generating dominant phenotypes. Development 1993;118:401–415.

9 Haigo S, Hildebrand J, Harland R, Wallingford J: Shroom induces apical constriction and is required for hingepoint formation during neural tube closure. Curr Biol 2003;13:2125–2137.

10 Kam Z, Minden J, Agard D, Sedat J, Leptin M: *Drosophila* gastrulation: analysis of cell shape changes in living embryos by three-dimensional fluorescence microscopy. Development 1991;112: 365–370.

11 Sweeton D, Parks S, Costa M, Weichaus E: Gastrulation in *Drosophila*, the formation of the ventral furrow and posterior midgut invaginations. Development 1991;112:775–789.

12 Myat MM, Andrew DJ: Fork head prevents apoptosis and promotes cell shape change during formation of the *Drosophila* salivary glands. Development 2000;127:4217–4226.

13 Chandrasekaran V, Beckendorf S: *senseless* is necessary for the survival of embryonic salivary glands in *Drosophila*. Development 2003;130:4719–4728.

14 Myat MM, Andrew DJ: Organ shape in the *Drosophila* salivary gland is controlled by regulated, sequential internalization of the primordia. Development 2000;127:679–691.

15 Xu N, Keung B, Myat MM: Rho GTPase controls invagination and cohesive migration of the *Drosophila* salivary gland through Crumbs and Rho-kinase. Dev Biol 2008;321:88–100.

16 Kolesnikov T, Beckendorf S: 18 wheeler regulates apical constriction of salivary gland cells via the Rho-GTPase-signaling pathway. Dev Biol 2007;307: 53–61.

17 Chandrasekaran V, Beckendorf S: Tec29 controls actin remodeling and endoreplication during invagination of the *Drosophila* embryonic salivary glands. Development 2005:3515–3524.

18 Smith A, Orr-Weaver T: The regulation of the cell-cycle during *Drosophila* embryogenesis: the transition to polytene. Development 1991;112:997–1008.

19 Myat MM, Andrew DJ: Epithelial tube morphology is determined by the polarized growth and delivery of apical membrane. Cell 2002;111:879–891.

20 Mosley-Bishop KL, Li Q, Patterson L, Fischer JA: Molecular analysis of the *klarsicht* gene and its role in nuclear migration within differentiating cells of the *Drosophila* eye. Curr Biol 1999;9:1211–1220.

21 Tepass U, Theres C, Knust E: *crumbs* encodes an EGF-like protein expressed on apical membranes of *Drosophila* epithelial cells and required for organization of epithelia. Cell 1990;61:787–799.

22 Tepass U, Knust E: Crumbs and Stardust act in a genetic pathway that controls the organization of epithelia in *Drosophila melanogaster*. Dev Biol 1993; 159:311–326.

23 Wodarz A, Hinz U, Engelbert M, Knust E: Expression of Crumbs confers apical character on plasma membrane domains of ectodermal epithelia of *Drosophila*. Cell 1995;82:67–76.

24 Kerman B, Chesire A, Myat MM, Andrew D: Ribbon modulates apical membrane during tube elongation through Crumbs and Moesin. Dev Biol 2008;320: 278–288.

25 Cheshire A, Kerman B, Zipfel W, Spector A, Andrew D: Kinetic and mechanical analysis of live tube morphogenesis. Dev Dyn 2008:2874–2888.

26 Pirraglia C, Jattani R, Myat MM: Rac GTPase in epithelial tube morphogenesis. Dev Biol 2006;290:435–446.

27 Castanon I, Von Stetina S, Kass J, Baylies M: Dimerization partners determine the activity of the Twist bHLH protein during *Drosophila* mesoderm development. Development 2001;128:3145–3159.

28 Vining MS, Bradley PL, Comeaux CA, Andrew DJ: Organ positioning in *Drosophila* requires complex tissue-tissue interactions. Dev Biol 2005;287:19–34.

29 Bradley PL, Myat MM, Comeaux CA, Andrew DJ: Posterior migration of the salivary gland requires an intact visceral mesoderm and integrin function. Dev Biol 2003;257:249–262.

30 Jattani R, Patel U, Kerman B, Myat MM: Deficiency screen identifies a novel role for beta 2 tubulin in salivary gland and myoblast migration in the *Drosophila* embryo. Dev Dyn 2009;238:853–863.

31 Kolesnikov T, Beckendorf S: Netrin and Slit guide salivary gland migration. Dev Biol 2005;284:102–111.

32 Rothberg J, Jacobs J, Goodman C, Artavanis-Tsakonas S: Slit: an extracellular protein necessary for development of the midline glia and commissural axon pathway contains both EGF and LRR domains. Genes Dev 1990;4:2169–2187.

33 Kidd T, Bland K, Goodman C: Slit is the midline repellent for the Robo receptor in *Drosophila*. Cell 2001;96:785–794.

34 Sato M, Umetsu D, Murakami S, Yasugi T, Tabata T: DWnt4 regulates the dorsoventral specificity of retinal projections in the *Drosophila melanogaster* visual system. Nat Neurosci 2005;9:67–75.

35 Peifer M, Polakis P: Wnt signaling in oncogenesis and embryogenesis – a look outside the nucleus. Science 2000;287:1606–1609.
36 Harris KE, Beckendorf SK: Different Wnt signals act through the Frizzled and RYK receptors during *Drosophila* salivary gland migration. Development 2007;134:2017–2025.
37 Yoshikawa S, McKinnon R, Kokel M, Thomas J: Wnt-mediated axon guidance via the *Drosophila* Derailed receptor. Nature 2003;422:583–588.
38 Duchek P, Somogyi K, Jekely G, Beccari S, Rorth P: Guidance of cell migration by the *Drosophila* PDGF/VEGF receptor. Cell 2001;108:865–876.
39 Harris KE, Beckendorf SK: Two ligands signal through the *Drosophila* PDGF/VEGF receptor to ensure proper salivary gland positioning. Mech Dev 2007;124:441–448.
40 Myat MM, Lightfoot H, Wang P, Andrew D: A molecular link between FGF and Dpp signaling in branch-specific migration of the *Drosophila* trachea. Dev Biol 2005;281:38–52.
41 Seshaiah P, Miller B, Myat MM, Andrew DJ: Pasilla, the *Drosophila* homologue of the human Nova-1 and Nova-2 proteins, is required for normal secretion in the salivary gland. Dev Biol 2001;239:309–322.
42 Maybeck V, Roper K: A targeted gain-of-function screen identifies genes affecting salivary gland morphogenesis/tubulogenesis in *Drosophila*. Genetics 2009:543–565.

Monn Monn Myat
Weill Medical College of Cornell University
1300 York Avenue, New York, NY 10065 (USA)
Tel. +1 212 746 1246, Fax +1 212 746 8175
E-Mail mmm2005@med.cornell.edu

Tucker AS, Miletich I (eds): Salivary Glands. Development, Adaptations and Disease.
Front Oral Biol. Basel, Karger, 2010, vol 14, pp 48–77

Extracellular Matrix and Growth Factors in Salivary Gland Development

Sharon J. Sequeira · Melinda Larsen · Tiffany DeVine

Department of Biological Sciences, University at Albany, SUNY, Albany, N.Y., USA

Abstract

The interstitial extracellular matrix (ECM) and epithelial-cell associated basement membrane (BM) play critical roles in the morphogenesis and differentiation of developing salivary glands. Early studies used ex vivo organ culture and tissue recombination methods to identify the importance of the ECM in organ development. Incorporation of transgenic mice and molecular tools has facilitated progress in our understanding of the mechanisms by which ECM proteins influence SMG development. Recent work has identified alterations in the ECM, BM, and associated proteins in salivary gland diseases, including Sjögren's syndrome and salivary gland cancers, but the significance of such changes is not known. Understanding the basic mechanisms controlling morphogenesis and differentiation in mammalian organ development is the first step towards understanding pathogenesis. Molecular characterization of the function of the ECM and BM in cellular processes is critical for future development of therapeutic approaches in regenerative medicine and tissue engineering. Here we provide a historical overview of experiments defining the functions of the ECM, ECM receptors, and associated molecules in salivary gland development. We include a discussion of the function of ECM-associated proteases and major growth factor signaling components that are in some way regulated by the ECM or associated molecules. We conclude with a discussion of defects in ECM and BM occurring in salivary gland pathologies and speculation on future areas of research pertaining to further understanding of the function of the ECM in the salivary gland.

The submandibular, or submaxillary, salivary gland (SMG) initiates from a thickening in the oral epithelium that grows into the neighboring mesenchymal, or stromal, tissue. Once this epithelium develops into a primary bud attached to an elongated stalk, the process of branching morphogenesis begins, which ultimately results in the complex arborized structure that is the adult gland [reviewed in 1, 2]. The SMG has long been studied as a model system for the study of branching morphogenesis since the developing SMG can be removed from the embryo as early as embryonic day 12 (E12), with the day of plug discovery defined as day 0, and studied ex vivo as an organ culture [3, 4]. The process of cellular cytodifferentiation, whereby the cells take on their ultimate shape and function, occurs during the later stages of morphogenesis through independent mechanisms that also require the influence of extracellular matrix (ECM) proteins [5]. Studies performed over the past 60 years have provided insight into the mechanisms whereby the stromal-derived ECM and the specialized epithelial cell-associated ECM, the basement membrane (BM), influence the growth, morphogenesis, and differentiation of the SMG tissues.

The ECM is a network of macromolecules located within the stromal connective tissue compartment of organs that is constructed by the stromal cells. In the SMG, the ECM is a significant component of the developing organ, but it accounts for a diminishing volume of the tissue in the adult gland. The ECM is composed of networks of fibrillar proteins, network accessory proteins, hydrophilic heteropolysaccharides (glycosaminoglycans) that are either not attached (e.g. hyaluronan) or are attached to proteins (proteoglycans), growth regulatory proteins that are sequestered in the network, and proteases and their inhibitors that regulate cleavage of ECM and associated proteins. The structure of the ECM gives it strength and rigidity, in addition to multiple other functions. The ECM is tightly linked to cell membranes through direct attachment via cell-surface receptor proteins (i.e. integrins, and other transmembrane proteins) and, through these signal-transducing transmembrane receptors, directs cellular responses, including cell differentiation, migration, and polarization. Signals initiated from inside the cell cytoplasm are also transmitted through these transmembrane receptor proteins to the ECM, resulting in its modification and its dynamic remodeling [6, 7].

The BM is a specialized form of ECM, located directly adjacent to the epithelium that is primarily synthesized by these cells and influences both morphogenesis and cell differentiation. BMs are generally composed of two thin sheets of matrix proteins: a layer known as the lamina densa (30–70 nm) and an underlying layer of reticular fibers (3 nm), which are collectively known as the BM [8, 9]. Scanning electron microscopic studies of developing rat SMG suggested that the two layers are significantly thicker than this: 100–400 and 30–40 nm, respectively at the E16 stage [10]. Early studies documented the importance of the BM in maintaining the structure of the epithelial lobules and implied that the BM participates in regulating the process of branching morphogenesis [11–15] and in facilitating exocrine secretion by the acinar structures [16]. The BM is a dynamic structure that undergoes remodeling during salivary gland morphogenesis and differentiation [17–19] and participates in mediating changes in tissue shape during morphogenesis [20].

Many ECM molecules are known to play an active role in salivary gland morphogenesis and differentiation, and their specific functions will be discussed here. These functions are summarized in figure 1. Collagens [21] are an important component of the ECM. The BM includes laminins [22, 23], collagen type IV [24], and nidogen [25, 26]. Glycosaminoglycans [11], proteoglycans [12], and heparan sulfate proteoglycans, such as perlecan [27, 28], are also integral components of the BM (table 1).

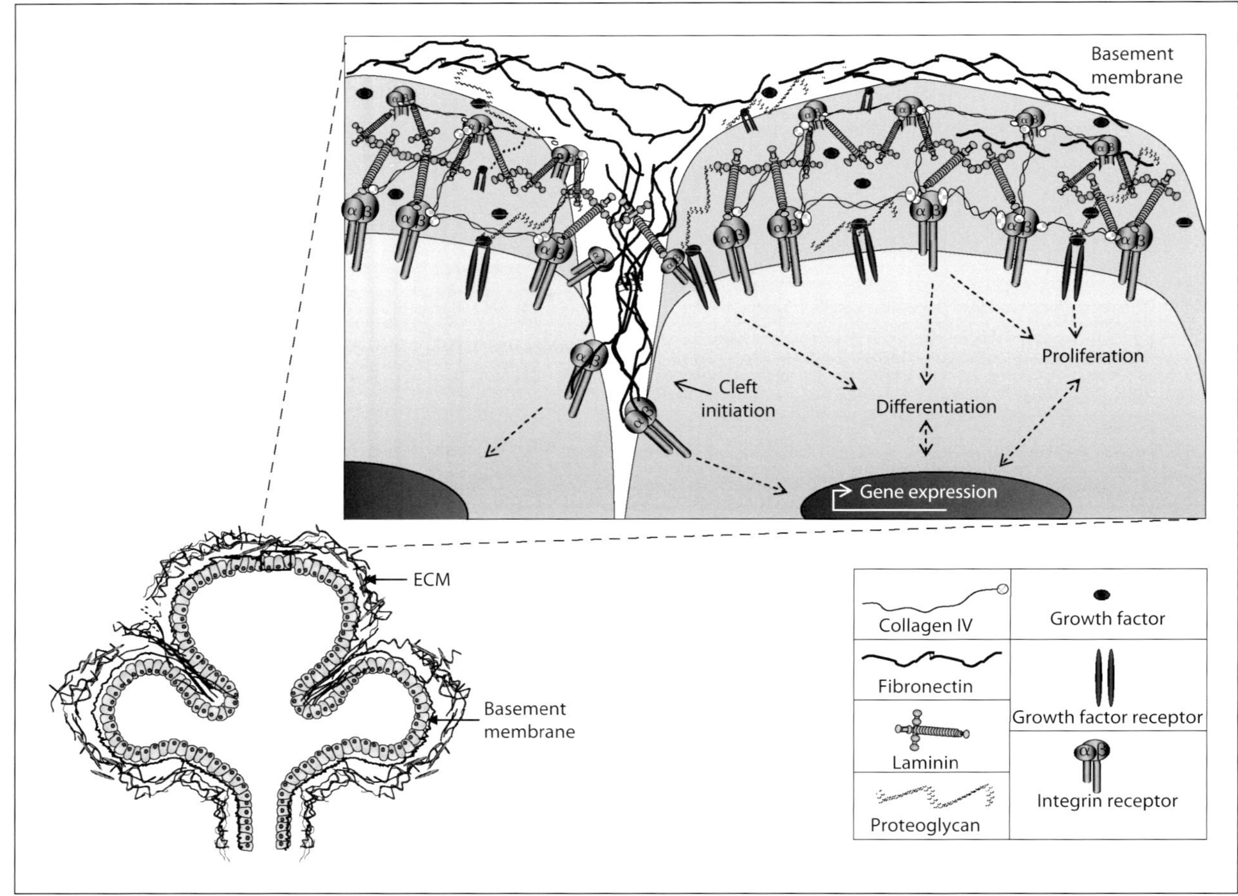

Fig. 1. Schematic of ECM-cell interactions during salivary gland development.

Fibronectin is a component of both the stromal ECM and the BM that is produced both by the mesenchyme [22] and by the epithelium and is critical to morphogenesis [29]. The ECM also serves as a storehouse for growth factors and as a mediator of growth factor interactions with their receptors. Proteases and protease inhibitors are involved in the release of growth factors and growth regulatory fragments of ECM proteins. Here we focus on studies that have impacted our understanding of the diverse functions that the ECM, BM, and associated molecules perform during SMG development. Although the other salivary glands (parotid, sublingual, and minor salivary glands) develop similarly, this discussion will focus primarily on the SMG. The majority of the studies mentioned were performed using mouse or rat cells and tissue, but most of the mechanisms discussed are likely to be conserved during human development.

Table 1. Summary of ECM and BM proteins involved in submandibular gland development

ECM molecule	Protein structure	Function in SMG development	References
Collagen			21, 42, 44, 45
Collagen I	triple helix, fibrils	major component of ECM	20, 24, 46, 48
Collagen III	triple helix, fibrils	cleft formation and/or maintenance	20, 24, 29, 49
Collagen IV	reticular fibers, BM	acinar morphogenesis and differentiation, branching morphogenesis	50, 51, 69
Laminin			23, 69
Laminin-111	trimeric ECM glycoprotein	acinar formation in vitro, cell survival, proliferation	60, 123
Laminin-α_1	laminin chain in trimer	binds syndecans, dystroglycans and integrins; spreading adhesion	56, 57, 59, 60, 61, 62
Laminin-α_5	laminin chain in trimer	lumen formation, regulates FGFR signaling	59, 63, 64, 65
Laminin-γ_1	laminin chain in trimer	BM formation, integrity, binds nidogen	25
Nidogen	globular glycoprotein	Linker molecule for BM proteins, laminin binding promotes morphogenesis	25, 60
Epimorphin	dual topology membrane protein	no known function in SMG development	59, 79
Tenascin C	ECM glycoprotein containing EGF- and FN-like repeats	potential role in ductal differentiation	81, 82
Proteoglycans	core ECM protein + GAG chains	branching morphogenesis, growth factor reservoir	84, 85
Perlecan/ heparan sulfate	core ECM protein + three GAG chains	connects ECM proteins to the BM, regulates FGF-FGFR binding affinity	27, 88, 89, 90, 95
Chondroitin-sulfate proteoglycan	sulfated GAG chains attached to core protein	branching morphogenesis, but not differentiation	14, 84, 97, 99, 100
Fibronectin	ECM glycoprotein dimer, assembles into fibrils	cleft formation and maintenance, cell proliferation	19, 24, 29, 107

Three-Dimensional Cell and Tissue Models

Our understanding of the function of ECM proteins in the developmental process of branching morphogenesis has greatly benefitted from ex vivo organ culture. In this technique, intact organs and tissues are harvested from live embryonic or adult mouse tissue and cultured in vitro for a finite period of time, during which the initial three-dimensional (3D) structure is maintained (fig. 2) [3, 4]. Borghese [30] was the first to realize that the mesenchymal tissue compartment was critical for the development of the epithelial tissue. Studies using tissue recombination models, in which epithelium from one tissue source is combined with the mesenchyme from another and then grown in a permissive in vitro environment, have demonstrated the inductive influence of the surrounding mesenchymal tissue and its accompanying ECM on the epithelial phenotype. The instructive function of SMG mesenchyme is clear from experiments in which it was combined with pituitary epithelium. The resulting epithelium within this tissue hybrid not only shows epithelial morphology, but also produces the salivary product, α-amylase, indicative of acinar differentiation [31]. The tissue recombination technique as performed here did not elucidate which specific molecules were involved. However, in many experiments, a Transwell filter placed between the epithelial and mesenchymal tissue did not negate the instructive function of the mesenchyme [32], indicating that the functional molecule(s) may be a soluble factor or a factor that can physically connect with the tissue through the pores of the filter. Studies later predicted that a soluble factor was necessary since the signals could be transmitted through a pore size as small as 0.05 μm [33].

In the embryonic salivary gland model, SMG epithelial cells dissociated from the surrounding mesenchyme will not grow alone. However, they can continue to undergo development if co-cultured with mesenchyme cells, implanted within an ECM gel and co-cultured with mesenchyme cells [33], or implanted in an ECM gel supplemented with mesenchymal-derived growth factors, such as fibroblast growth factors (FGFs) or epidermal growth factor (EGF) [34, 35]. The first ECM extract to be produced, which is still the most frequently used extract for 3D culture, was a BM extract derived from the Engelbreth-Holm-Swarm tumor that is known as Matrigel [36, 37]. Matrigel comprises a complex mixture of BM components, primarily: laminin-1, collagen IV, perlecan, and nidogen, and other minor components including growth factors and proteases [38]. In such ex vivo organ culture systems, Matrigel, together with exogenous growth factors, can adequately replace the native mesenchymal stroma and successfully promote branching morphogenesis and differentiation of the salivary gland epithelium [33–35, 39]. Further, Matrigel has been used to illuminate the processes of acinus formation and cytodifferentiation using salivary gland cell lines grown in a 3D format, such as the human neoplastic SMG intercalated duct cell line, HSG [40, 41].

In recent years it has been possible to use organ culture experiments in combination with pharmacological inhibitors, function-blocking antibodies, fluorescent

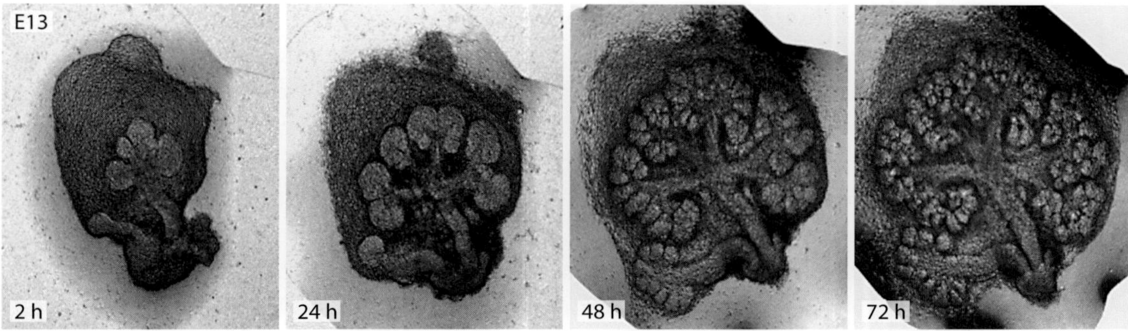

Fig. 2. Submandibular salivary gland growth ex vivo. E13 mouse SMGs were placed on polycarbonate filters floating on top of serum-free media and cultured for 72 h. Brightfield images were captured at 24-hour intervals. SMGs continue to grow and undergo branching morphogenesis during this 72-hour period, recapitulating in vivo development.

probes of various types, and various inhibitory RNAs (antisense and small inhibitory RNAs) to probe the functions of specific ECM molecules. Use of the intact SMG cultures has been compared with growth of epithelial rudiments grown in Matrigel, laminin, or other exogenous matrices, to answer specific questions regarding the role of the mesenchyme and ECM proteins. The combination of organ culture studies with studies of knockout mice has been a powerful combination for delineating molecular mechanisms controlling branching morphogenesis.

Collagen

It has been known for many decades that collagens play an important role in organogenesis. The most well-known collagens are the fibril-forming, or fibrillar, collagens that exist in rope-like fibers composed of multiple triple-helical structures primarily within the stromal compartment. Early studies revealed that collagen was important for the process of branching morphogenesis in SMG, since branching could be inhibited by treating mouse SMGs with collagenase [21]. Not only was further branching inhibited due to inability of SMGs to form clefts, or indentations in the BM at the epithelial surface, but existing clefts regressed. Subsequent studies showed that purified clostridial collagenase blocked new cleft formation but did not cause loss of established clefts despite the fact that it disrupted the continuous nature of the BM [42], suggesting that non-collagenase contaminants caused the loss of existing clefts observed in previous studies [43]. Spooner and Faubion [44] examined the function of collagen more specifically by using inhibitors of collagen synthesis (L-azetidine-2-carboxylic acid) and secretion (α,α-dipyridyl), both of which inhibited SMG morphogenesis. Conversely, treatment of SMGs with collagenase inhibitors significantly increased the number of clefts and the density of collagen fibrils found to localize specifically within clefts, as detected by scanning electron microscopy [42, 45]. While these studies together implicate fibrillar collagens in stabilizing clefts, these experiments did not identify the specific functional collagen isoform(s).

Treatment with bovine dental pulp collagenase, which targets collagen I and III, produced SMGs phenotypically similar to those treated with the less specific clostridial enzyme used in earlier studies [46]. Immunolocalization studies subsequently showed that collagen type III is found in the early narrow clefts of E13 SMGs [47], suggesting a function in branching morphogenesis. However, treatment of SMG organ cultures with increasing doses of collagen III had no measurable morphological effect on branching morphogenesis [29]. While organ culture data has strongly implicated several collagen isoforms in morphogenesis, knockout mouse models have so far contributed little to our understanding of the function of collagens in SMG development possibly because other collagen isoforms or alternative ECM components may compensate for them in their absence. The *collagen I (Mov-13)* null mouse appears to be normal and showed no defects in salivary gland development when glands were cultured ex vivo [48], however other collagens can compensate for collagen I. *Col3a1–/–* mice, on the other hand, displayed a significant number of other problems and had only a 10% survival rate to adulthood, yet a salivary gland phenotype was not reported [49]. In light of the knockout mouse data, additional studies are needed to clarify the function of fibrillar collagens during salivary gland development.

The salivary gland BM contains collagen type IV, which forms a flexible network providing both tensile strength and growth regulatory functions. Function-blocking collagen IV antibodies significantly inhibited branching morphogenesis of SMG organ cultures [50]. Although these SMGs showed relatively normal ductal differentiation, no terminal end buds formed, and secretory differentiation did not occur, suggesting that collagen IV is important for both acinar morphogenesis and differentiation. Recently, fragments of collagen IV, NC1 domains released by the matrix metalloprotease, MT2-MMP, have been shown to stimulate branching morphogenesis [51]. Acting through the ECM receptor, β_1 integrin, NC1 domains were shown to stimulate epithelial proliferation and branching morphogenesis in organ cultures, most likely though upregulation of HB-EGF and FGF1. Interestingly, *Col4a2* transcription increased when MT-MMP function was reduced, indicative of a complex interaction between MMPs and collagen IV in the SMG. Collagen IV appears to play a central role in integration of MMP-induced activities and growth factor-induced regulation of cell proliferation and morphogenesis.

Laminin

The laminin network forms an essential part of the BM, which is linked to other BM components, such as collagen IV, through linker proteins like nidogen and to the intracellular cytoskeleton through cell-surface receptors. Laminins directly interact with the outer, basal layer of epithelial cells in the SMG throughout development. Currently, 16 different laminin heterotrimers have been described in mammals, which are assembled from five different α, four β, and three γ chains [52, 53] and are named with a three-digit number for the chains they contain [54].

Through rotary shadowing microscopy, it is apparent that the laminin molecules physically resemble cross-like structures; the 'arms' of the cross carry out specific functions. The β and γ chains each form a short arm, which contain sites that regulate laminin self-assembly and interaction with other ECM proteins. Laminins are thought to initiate BM assembly and then nucleate assembly of other molecules [53]. A nidogen-binding site located within the short arm of the γ_1 chain is important for SMG branching morphogenesis [25]. The distal portion of the long arm is a coiled coil containing all three chains, with a group of five laminin globular (LG) G-domains located at the C-terminus of the α chain. These G-domains are implicated in multiple interactions with laminin cell-surface receptors, including at least eight integrins [55], syndecans [56], and dystroglycan [57].

Within the salivary gland, different laminin isoforms have distinct temporal distributions [58]. The laminin α_1 chain is highly expressed within all SMG BMs at the initiation of branching morphogenesis, but its expression progressively decreases during development and becomes restricted to ductal BMs in the adult [59]. The functions of the laminin-α_1 chain during the process of branching morphogenesis have been studied using active peptides and function-blocking antibodies in organ culture. Function-blocking laminin-α_1 antibodies perturb SMG branching morphogenesis [59]. The LG1–5 domains located within the C-terminus of the α chain have been extensively studied using these tools. AG-73, a synthetic peptide sequence represented within the LG4 module, which mediates cell spreading and adhesion, inhibits branching of E13 SMGs [60, 61]. Explants treated with AG-73 show evidence of abnormal BM formation, typified by breaks and thin irregular areas, suggesting that the laminin α_1 LG4 module is involved in its assembly [61]. The LG4 module has been identified as a site for both α-dystroglycan [57] and syndecan-1 binding [56], and antibodies blocking the interaction between α-dystroglycan and laminin-α_1 are capable of perturbing branching. These rudiments, which have large lobes that have failed to undergo clefting, morphologically resemble those treated with mAB 200, an antibody targeting the LG4 domain of α_1 [57]. Knockout mice lacking the α_1 chain die by E7 [62] and therefore cannot be analyzed using ex vivo SMGs – a conditional mutation model would have to be created. Surprisingly, BMs in the laminin α_1 knockout animals seem to form normally, suggesting a partial compensation by other laminins, although subsequent studies indicate that each isoform has specific functions.

The laminin α_5 chain, found in laminin-511 and laminin-521, is expressed early in salivary gland development and incorporated within the BM [63]. Expression of the α_5 chain increases during the early stages of branching as α_1 decreases [63, 64]. Treatment of explants with A5G77f, a peptide generated from the sequence of the LG4 module, identified a function for the laminin α_5 chain in SMG development that is distinct from laminin α_1. Explants treated with A5G77f did not show a smooth BM, but instead were covered with deep clefts, and abnormal contours were observed in the ducts [64]. The α_5 chain plays a critical role in the process of development, as evidenced by the phenotype of *Lama5* null mice. They survive only until E17 and show severe defects in multiple organs [65], including defective SMG development and a failure of the sublingual glands to develop in a significant number of cases [63, 64].

Subsequent studies utilizing a laminin α_5 knockout mouse identified a more specific role for α_5. Abnormal cytodifferentiation was noted by E15, absence of a lumen, and a high degree of cellular disorganization in the developing acini was observed [63], highlighting the importance of this α chain in later stages of SMG development. The *lama5* null mouse and siRNA knockdown of the message both showed a decrease in MAPK (p42) phosphorylation and expression of FGF1, FGF receptor 1b (FGFR1b), and FGFR2b [63]. Laminin α_5 signaling was shown to regulate FGFRs through β_1 integrin. Interestingly, FGFRs also regulate *lama5*. This pathway is an example of the close relationship between BM proteins and growth factor signaling pathways during salivary gland development.

Nidogen

Nidogens, also known as entactins, consist of a number of sulfated monomeric glycoproteins. Nidogen 1 (150 kDa) and nidogen 2 (200 kDa) are ubiquitously expressed BM proteins that consist of three globular domains (G1–G3) connected by thread and rod-like segments. Nidogens function as linker molecules and bind to a wide repertoire of other BM components, including laminins, collagen IV, perlecan, and fibulins [66]. The G2 domain binds to type IV collagen, perlecan (domain IV-1), fibulin-1, and fibulin-2, while the G3 domain can bind to laminin γ_1 and collagen IV; these interactions are thought to be particularly important for BM assembly. The G1 domain has moderate binding affinity for fibulin-2 [67]. Fibulins are secreted BM glycoproteins having multiple EGF-like domains that are known to mediate tissue remodeling [68] but have no demonstrated role in SMG development.

Nidogen-1 can be detected during mouse embryonic development as early as the morula stage [69], and it co-localizes with laminin in the BM of the developing mouse embryo [70]. However, neither nidogen 1- [71] nor nidogen 2-deficient mice [72] showed BM defects, suggesting possible compensatory functions for each protein. In contrast, nidogen double null mice die shortly after birth with various abnormalities in the lungs, heart, and limbs resulting from BM defects [73]. However, the finding that nidogen double null mice can, in many tissues, form BMs that appear to be normal by electron microscopy suggests that nidogens are not essential for formation of all BMs, and that BMs from each tissue have different requirements for nidogens.

Nidogen protein is present in the BM throughout pre- and postnatal developmental stages, and in situ analysis demonstrates that nidogen is made by the mesenchyme [25, 26]. Antibodies directed against the nidogen-binding site in the laminin γ_1 chain that disrupt this interaction reduced SMG branching morphogenesis [25]. Reconstituted laminin-111-nidogen gels can support SMG epithelial bud growth and branching morphogenesis [60], however laminin-111 gels lacking nidogen also support SMG epithelial branching morphogenesis [74]. Since both nidogen 1 [75] and nidogen 2 [76] bind to the third immunoglobulin module (IG3) in domain VI of the protein core of perlecan via the central G2 domain the large complexes formed by nidogen-perlecan-laminin interactions appear to be important to create the supramolecular organization of the SMG BM.

Epimorphin

Epimorphin (also called syntaxin-2), is a multifunctional mesenchymal cell-surface protein, which was postulated to be an epithelial morphogen in many embryonic tissues, including salivary, mammary, gallbladder, pancreas, lung and liver epithelial cells [reviewed in 77]. Epimorphin is incorporated into the BM of the primary salivary duct regions from E13 onwards during embryonic growth. However, function-blocking antibodies directed against epimorphin failed to block epithelial development of the embryonic SMG [59], despite the fact that epimorphin influences lung morphogenesis in organ culture assays [78]. Epimorphin null mice show defective testicular growth and enhanced intestinal growth but do not have a reported SMG phenotype [79]. If epimorphin plays a role in SMG development, it has yet to be identified.

Tenascin C

Tenascin C is another mesenchymal-expressed protein and is considered to be a 'matricellular protein', i.e. a protein that is incorporated into the ECM or BM and regulates cell behavior but is not a structural protein [80]. Its expression was detected adjacent to blood vessels in embryonic rat SMGs, but increased expression in acinar and ductal BMs was detected after birth [81]. Transgenic mice lacking the tenascin C gene do not have visible defects in organ development [82], and the distribution of fibronectin, laminin, collagen, and proteoglycans was not affected in these animals. In human embryonic salivary glands, tenascin was detected only near excretory ducts [83]. While the knockout phenotype indicates no function for tenascin C in SMG development, its expression pattern suggests a potential contribution towards ductal differentiation.

Proteoglycans

The classic proteoglycans (PGs) are made up of glycosaminoglycan (GAG) chains that are bound by glycosidic linkages to a core peptide. The functions of PGs in salivary glands are mediated via their GAG chains since prevention of GAG assembly on the protein core of PGs inhibited salivary gland branching morphogenesis [84, 85]. Such methods also indicated that chondroitin sulfate was more important than heparan sulfate PG in SMG branching morphogenesis [85, 86]. The GAG chains can act as reservoirs for various growth factors, e.g. basic FGF/FGF2 or FGF7 as demonstrated using cell lines and reviewed in Ruoslahti and Yamaguchi [87], and thus can potentially influence morphogenesis by more than one mechanism.

Recent studies have identified important roles for heparan sulfate in regulation of growth factor activity in SMGs [88, 89]. Heparanase, an endoglycosidase that degrades heparan sulfate, released FGF10 from its attachment to perlecan heparan sulfate in the BM, thereby increasing epithelial clefting and lateral branch formation to promote branching morphogenesis [88]. Addition of heparan sulfate to FGF10-stimulated epithelial rudiments increased both proliferation and expansion of end buds. Furthermore, specific modification of the heparin with 2-O-sulfation with either an N- or 6-O-sulfate induced end bud expansion and also an increased expression of end bud differentiation markers. Conversely, heparin decasaccharides modified with 6-O-sulfation alone induced duct elongation along with a ductal differentiation marker [88–90]. While this study demonstrated that the specific sulfation patterns of heparin decasaccharides alter intracellular signaling, interestingly regulation of differentiation and morphogenesis appear to be regulated coordinately in this case, in contrast to previous findings suggesting that in general that differentiation and morphogenesis are not coupled [5].

Perlecan (heparan sulfate proteoglycan 2, HSPG2) is a BM HSPG with a wide tissue distribution and multiple functions during organogenesis [27]. The GAG chains, located in the N-terminal domain of the core protein, bind multiple growth factors and also interact with other BM components, such as nidogens, laminins [75] and collagen IV [91]. The 400- to 450-kDa core protein, composed of several protein modules arranged in five distinct domains, has multiple binding partners including FGF-7, fibronectin, heparin, laminin-1, platelet-derived growth factor (PDGF-BB), and integrins [reviewed in 92]. Interestingly, the perlecan domain IV domain within the core protein appears to be a binding 'hotspot' in which its immunoglobulin-like domains (IG), IG2, IG4, and IG5, interact with fibulin-2, fibronectin, and heparin, respectively [91]. Perlecan is expressed very early during the blastocyst stage and is critical for embryonic development, as evidenced by studies using perlecan knockout mice. No abnormalities are apparent before E10, but most embryos die between E10 and E12, with evidence of bleeding into the pericardial sac; however, animals that survive to birth die shortly afterwards due to severe brain and skeletal defects. Interestingly, most BMs appear to be normal in homozygous embryos [27], suggesting that either perlecan may not have a critical role in BM assembly or that other PGs like agrin, which shares homology with perlecan, or collagen XVIII, may substitute for it [93, 94].

Perlecan also participates in signaling through more than one mechanism. Perlecan regulates the activity of FGFs by acting as a co-receptor, enhancing FGF-FGFR affinity, and serving as a storage reservoir for FGFs in the ECM [95]. Inhibition of heparanase, an endoglycosidase which cleaves FGF-FGFR complexes bound to perlecan, decreased SMG branching morphogenesis while exogenous heparanase increased epithelial clefting and SMG branching [88]. Purified perlecan domain IV peptides have also been shown to support the growth and cytodifferentiation of dissociated human adult salivary gland acinar cells in vitro. These cells are able to organize adherens junction proteins like E-cadherin and maintain their differentiated status, as evidenced by expression of aquaporin 5 and α-amylase activity in the presence of the peptides [96].

Chrondroitin sulfate proteoglycans (CSPG) are the predominant GAGs produced during early SMG development. As the gland matures, levels of this PG significantly decrease while those of HSPG increases [14, 50, 97, 98]. CSPG was shown to be required for embryonic salivary gland branching morphogenesis but not secretory cell differentiation since E16 rat embryonic SMGs treated with β-D-xyloside, which uncouples GAG chains from the PG core protein, showed deregulated CS synthesis and failed to branch but were able to synthesize secretory granules [84, 85, 99]. In contrast, addition of exogenous PG to SMG epithelial rudiments did not affect branching morphogenesis as did removal of the heparan sulfate chains of heparan sulfate proteoglycans (HSPGs) with heparitinase [100]. The mechanistic role of chondroitin sulfate proteogylcans in early salivary gland development warrants further analyses.

There are other GAGs that may be important in SMG branching morphogenesis that have not been studied in depth. The leucine-rich proteoglycan (SLRP) family consists of a protein core containing leucine repeats with a GAG chain consisting of either CS or dermatan sulfate. The possible function of this family in SMG development is unknown. Decorin is another matrix PG closely related in structure to biglycan that binds to type I collagen and is involved in ECM matrix assembly. Although decorin was upregulated at the mRNA level by adhesion to FN or Collagen I in HSG cells [101], its function in SMG development has not been determined, nor has a function been identified for other family members (biglycans or agrin) in SMG development.

Fibronectin

Fibronectin (FN) is a glycoprotein that can exist as many variants due to alternative splicing. Although hepatocytes secrete soluble plasma FN into the bloodstream where it is readily recruited to wound repair sites, FN is also synthesized and secreted by many tissues where it is most commonly incorporated into the ECM, but it is also incorporated into many developing BMs [22] in a fibrillar cell-associated form. FN is secreted as disulfide-bonded dimers that are assembled into cell-associated fibrils through a complicated process involving integrin-mediated stretching of the fibrils to expose cryptic sites in FN needed for self-assembly [reviewed in 102]. The FN network is integrally associated with other ECM proteins through specific binding sites within the molecule. It is directly bound to both collagens and cell-surface receptors, while indirectly linked to laminin through nidogen [reviewed in 103]. While integrin $\alpha_5\beta_1$ is the primary cell-surface receptor for FN [104], other integrins and syndecans can also function as receptors [105] to mediate contact with the intracellular actin cytoskeleton and initiate intracellular signaling. The fibrillar forms of FN are critical for many essential cellular events required during development, including adhesion, migration, spreading and growth. In fact, mice lacking FN show severe developmental defects that ultimately result in embryonic lethality at E9 [106], so it has not been possible to examine the function of FN in development of branching organs using this mouse model.

In mouse SMGs, FN-mediated cell-matrix adhesions are critical to the process of branching morphogenesis [29]. FN, shown by in situ hybridization to be produced by the epithelium during the initial stages of branching, localizes adjacent to newly forming clefts. siRNA knockdown of FN decreased branching by inhibition of cleft formation, whereas the addition of exogenous FN rescued this phenotype in a dose-dependent manner, and cleft formation resumed [29]. Insight into the mechanism by which FN is involved in cleft formation and progression was provided through use of confocal time-lapse microscopy. Pulse-chase experiments with fluorescently-labeled FN revealed that older FN forms a wedge at the most central point of progressing clefts, which was suggested to externally stabilize clefts among constantly shifting immature epithelial cells during cleft formation [19]. It was also reported that FN accumulation at clefts was associated with decreased E-cadherin expression, suggesting a negative regulation of cell-cell adhesions and replacement with cell-ECM adhesions within the cleft (fig. 1) [29]. Recent work has demonstrated that FN assembly occurs downstream of Rho-kinase (ROCK) and myosin-mediated assembly during SMG branching morphogenesis. ROCK signaling regulates a mechanochemical checkpoint that transitions initiated clefts into clefts competent for progression [107]. FN assembly in the cleft was shown to be downstream of this checkpoint and to stimulate cell proliferation near the bud periphery during cleft progression. Whether FN is the key BM component coordinating BM assembly during SMG cleft formation remains to be determined.

Integrins

The integrins are a major family of transmembrane receptors that connect cells to the ECM through their extracellular domain, and anchor cytoskeletal proteins to the plasma membrane through their intracellular domain. There are currently more than 20 integrin receptors, formed by heterodimerization of α and β subunits. Each of the major ECM components can be recognized by one or more integrins [reviewed in 7, 13, 108]. That integrins are important in salivary gland development was demonstrated by the finding that integrins are required for the posterior migration of the *Drosophila* salivary gland during development [109]. Normal mammalian salivary gland epithelial cells have been known to express the VLA integrins ($\alpha_1\beta_1$-$\alpha_6\beta_1$), and β_4 integrin subunits for many years [110–113]. The importance of β_1 integrin in both human and mouse salivary gland development is highlighted by studies showing that β_1 expression increases during the branching and initial canalization and cytodifferentiation stage of human salivary gland development [114] and by studies using anti-β_1 function-blocking antibodies that significantly inhibited mouse embryonic SMG branching, cleft formation [29], and cell migration [29, 115] (table 2). Multiple integrin α subunits can partner with the β_1 subunit to elicit specialized functions in the salivary gland.

The first integrin identified to play an important role in SMG branching morphogenesis was integrin α_6. Integrin α_6 can associate with either β_1 or β_4 to form laminin-binding heterodimers. Integrin $\alpha_6\beta_1$ functions as a major laminin receptor in embryonic SMG, and function blocking antibodies to integrin α_6 (GoH$_3$) significantly blocked salivary gland branching morphogenesis [59]. Further, function-blocking anti-α_6 antibodies also affected E-cadherin distribution in the inner cell mass of epithelial buds in SMGs [59]. The primary FN receptor, integrin $\alpha_5\beta_1$ was found to be critical for SMG branching morphogenesis [29]. Antibodies against the α_5 subunit substantially decreased branching in a dose-dependent manner [29]. The combination of antibodies targeting integrin α_5 and integrin α_6 inhibited branching more effectively than each antibody separately and at a level approximately equal to inhibition with anti-integrin β_1 antibodies, suggesting that both laminin and fibronectin are important in branching [29]. Several collagen-binding integrins are known, including $\alpha_1\beta_1$, $\alpha_2\beta_1$, $\alpha_3\beta_1$, $\alpha_{10}\beta_1$ and $\alpha_{11}\beta_1$; however, whether these integrin heterodimers bind to collagen in the salivary gland ECM and participate in SMG development and function is not yet understood. Gene and protein expression studies using HSG epithelial cells adhered via integrin receptors to different ECM substrates, such as collagen I or fibronectin, yielded more than 30 differentially-expressed genes [101]. Thus further characterization of integrin-mediated changes in the levels of adhesion-responsive genes is warranted.

Altered integrin expression can also lead to changes in cell polarity which may contribute to salivary gland disease phenotypes. Mice carrying a targeted mutation in the α_3-integrin gene, *itga3*, develop aberrantly differentiated SMGs with defects in apico-basal polarity and in the BM. Integrin α_3-deficient SMGs exhibited altered expression and/or localization of several ECM and adhesion molecules, including laminin, fibronectin, integrin β_1, integrin α_5, and E-cadherin [116]. Given the recently identified altered expression of integrins and ECM proteins in Sjögren's syndrome,

Table 2. Summary of integrin receptors involved in salivary gland development

Integrin	ECM ligand	Function in salivary gland	References
β_1	multiple	SMG branching, cleft formation, migration	29, 114, 115
$\alpha_2\beta_1$	collagen, laminins	altered expression in Sjögren's syndrome	161
$\alpha_3\beta_1$	laminins	differentiation and apico-basal polarity	116
$\alpha_5\beta_1$	fibronectin	branching morphogenesis, cleft formation	19, 29
$\alpha_6\beta_1$	laminins	differentiation	29, 59
$\alpha_6\beta_4$	laminins	laminin receptor, differentiation	59, 108, 110

mechanistic studies regarding integrin-mediated signaling during development are of interest.

Non-Integrin ECM Receptors

Little is understood regarding the role of non-integrin receptors in salivary gland development and function. However, given the essential roles of ECM molecules such as laminins, it is reasonable to assume that non-integrin laminin-binding receptors may contribute significantly to embryonic salivary gland development.

Dystroglycan is a receptor for multiple ECM molecules, including laminin-111, laminin-211, perlecan and agrin, and links the ECM to the actin cytoskeleton [117]. Branching morphogenesis of embryonic kidneys is inhibited when organs are cultured in the presence of anti-α-dystroglycan antibodies, which perturb its binding to laminin-111 [118]. Dystroglycan mRNA is expressed in the epithelium of developing salivary glands and lung. Further, antibodies against dystroglycan can inhibit branching morphogenesis in salivary gland and lung organ cultures [57]. Other non-integrin receptors such as the 67-kDa laminin receptor, 67LR, are expressed in malignant salivary gland neoplasms, such as adenoid cystic carcinomas. 67LR binds a sequence of laminin β_1 chain (YIGSR) and contributes to the transformed phenotype in adenoid cystic carcinoma cells cultured in 3D [119].

Syndecan-1 (sdc-1 or HSPG1) is a member of a family of cell-surface proteoglycans that bind cells to ECM molecules [120]. It also binds FGFs, may act as a co-receptor to enhance FGF binding to its receptor, and serves as a storage reservoir for FGFs in the ECM [88]. Mice lacking syndecan-1 are morphologically normal but

Table 3. Summary of proteases and protease inhibitors involved in salivary gland development and disease

Protein	Protein structure	Function in SMG	References
MMP1	secreted metalloprotease	cleaves collagen	42, 46
MMP2	secreted metalloprotease	ECM remodeling	74, 146
MT1-MMP	transmembrane metalloprotease	activates MMP2	74, 146
MT2-MMP	transmembrane metalloprotease	releases bioactive collagen IV NC1 domains	51
TIMP	secreted inhibitor	inhibits MMP activity	90, 164, 165

exhibit delayed skin and corneal wound healing and defective regulation of integrin α_9 expression [121]. Sdc-1 functions to regulate integrin expression levels mediated through transforming growth factor (TGF)-β_1 signaling to modulate cell adhesion and migration in response to ECM substrates, such as fibronectin, collagen I, or laminin-332 in keratinocytes [122]. Sdc-1 acts as a receptor for the laminin α_1 G4 domain peptide, AG73, in HSG cells [123]. Sdc-1 is expressed in all epithelial cells in developing SMGs [124], yet its functions in SMG development remain to be determined.

Growth Factor Signaling

The function of growth factors and their interaction with their receptors can be modified by the ECM and associated molecules by many mechanisms. The ECM can act as a storage reservoir for growth factors that are restrained from free diffusion due to the ECM. Heparan sulfate and similar ECM molecules can modify the activities of growth factors by various mechanisms. Proteases, particularly those in the matrix metalloproteinase (MMP) family, in addition to releasing bioactive fragments of the ECM (see table 3) have been implicated in activating growth factors from their inactive form.

Fibroblast Growth Factors. The FGF family of growth factors, which is comprised of at least 22 members and directs diverse biological functions, including cell proliferation, branching morphogenesis, and differentiation, by selectively binding to and activating four tyrosine kinase receptors (FGFRs 1–4) [125–127]. Expression of the FGF ligands and receptors is spatiotemporally regulated during salivary gland development [74]. FGF10 and its receptor, FGFR2, isoform IIIb (FGFR2b) [128, 129], are known to be critical for mouse SMG development. This was discovered by studies showing that FGF10 and FGFR2b null mice [130–132] display SMG aplasia while FGF10 heterozygous mice are viable and fertile but display both SMG and lacrimal gland hypoplasia [133]. FGFR2b-mediated signaling serves to integrate both ECM-dependent (laminin-511) as well as integrin-dependent (β_1) signals, to control SMG proliferation epithelial clefting and branching [63, 74]. FGF-FGFR signaling is enhanced by the presence of heparan sulfate in the ECM, which can protect FGFs from proteolytic degradation and facilitates FGF internalization, processing, and release from HSPGs like perlecan or sdc-1 [88]. Since FGF7 also binds to FGFR2b, but the FGF7 knockout mouse does not show a defect in SMG development, it was puzzling as to why FGF7 and FGF10 promote branching and ductal morphologies, respectively. In a recent paper, Makarenkova et al. [134] demonstrate that the key difference may lie in their heparan sulfate-binding strength and ability to diffuse through the ECM. FGF10 binds more strongly to heparan sulfate and shows enhanced diffusivity within the ECM, relative to FGF7. In an elegant set of experiments, in which single amino acids were mutated in FGF10, the authors demonstrate that a mutation preventing heparan sulfate binding and diffusion through the ECM can change the response of SMG epithelial rudiments to the mutant FGF10 to mirror their response to FGF7, unlike mutations that only affect receptor binding.

Epidermal Growth Factors. EGF family members are intricately related to ECM production in the SMG and are also regulated by the ECM. EGF, TGF-α and EGFR, which functions as a receptor for both ligands, are all expressed during early salivary gland development [135], and EGF can support growth and branching morphogenesis of mouse embryonic SMG explants in in vitro organ cultures [33, 34]. EGF is important for facilitating interactions between salivary epithelial cell receptors and the surrounding ECM. For example, addition of EGF into the medium of salivary gland organ cultures, increased nidogen-1 expression and could rescue SMG explants from the effects of antibodies that inhibited high-affinity nidogen-laminin binding [25]. Signaling via the EGF receptor (EGFR), upon ligand binding, was shown to be required for SMG development and to regulate the expression of integrin α_6, and thus, in turn, regulate epithelial-mesenchymal interactions in early gland development [136]. Heparin-binding EGF-like growth factor (HB-EGF) is abundantly expressed in the epithelium of the developing mouse salivary gland [90], more so than EGF. HB-EGF, which also signals through EGFR, is likely to be a more physiologically-relevant ligand for EGFR in SMG than EGF itself. Cleavage of the nascent HB-EGF membrane protein by matrix MMPs yields active soluble growth factor. Inhibition of HB-EGF activity using synthetic peptides corresponding to the heparin-binding domain of HB-EGF or anti-HB-EGF neutralizing antibodies inhibited SMG morphogenesis. Further, the inhibitory effects on morphogenesis were reversed, though partially, by adding exogenous soluble HB-EGF or by pharmacological inhibition of MMP activity [90]. The activity of HB-EGF is further regulated by interactions with HSPGs, which are required for EGFR activation [137].

Insulin Growth Factors (IGFs). The IGF family includes three ligands (insulin, IGF-I, and IGF-II) and three surface receptors (insulin receptor, IGF-IR, and IGF-IIR/mannose 6-phosphate receptors). IGF signaling is dependent upon at least six high-affinity extracellular IGF-binding proteins (IGFBPs), which bind circulating IGFs and control growth and differentiation in many oral tissues including teeth, mandibles, maxillae, and the tongue. IGF-I and II are present in human saliva and IGF-II has also been shown to be overexpressed in salivary gland adenomas, suggesting that aberrant IGF signaling may be a causative factor in the etiology of oral cancers [138]. Treatment of the HSG cell line with tumor necrosis factor-α and interferon-γ caused a significant decrease in cellular proliferation that was associated with increased accumulation of IGFBP-3. This effect on proliferation was shown to be dependent upon IGFBP-3 since it was reversible upon treatment with antibodies to IGFBP-3 [139]. Studies on osteoblastic cells and bone tissue have implicated a strong role for ECM regulation of IGFs and their binding partners. IGFBP-2 in its native form has little affinity for ECM derived from human or rat osteoblastic cells. However, in the presence of IGFs, IGFBP-2 binding to ECM was markedly enhanced. In the presence of IGF-II, IGFBP-2 bound with high affinity to heparin-sepharose, but not to type I collagen, fibronectin, or laminin. Furthermore, heparin and HSPG, but not CS, inhibited IGFBP-2/IGF-II binding to ECM, thus demonstrating selectivity of certain ECM molecules to facilitate IGF-mediated signaling [140]. High levels of IGF1 were reported in salivary gland [141–143] and expression of IGFII, IGF-IR and IGF-IIR was also reported. IGFII and IGFIIR signaling may also cross-talk with EGF signaling and participate in regulation of TGF-β release and activation in developing SMGs together with proteases [144].

Proteases and Protease Inhibitors

The importance of proteases in branching morphogenesis was first demonstrated in organ culture experiments using tissue inhibitors of matrix metalloproteases (TIMPs). The addition of TIMPs in culture promoted branching of SMG organ cultures by increasing the number of clefts [145], demonstrating the likely presence of endogenous matrix metalloproteases (MMPs) in SMGs. That MTI-MMP is important for salivary gland development was demonstrated using *mmp14* null mice. In addition to musculoskeletal abnormalities, these mice have severe salivary gland defects, and branching morphogenesis is inhibited when *mmp14* null SMGs are cultured ex vivo [146]. *Mmp14* encodes the transmembrane protein MT1-MMP, which acts in a signaling cascade to activate other MMPs, including MMP2 and MMP13. The severe phenotype presented by *mmp14* null mice was interpreted to be due primarily to the inactivation of downstream MMP2. In fact, Steinberg et al. [74] showed that MMP2 regulates FGF7-mediated SMG branching and cleaves FGFR1. The authors suggest that this cleavage product may compete for access to FGFs to regulate signaling. Although *mmp2* null mice appear to develop normally and lack the characteristic SMG defects associated with the *mmp14* null phenotype, this *mmp2–/–* phenotype may be due to developmental compensation by other MMPs, which is a common issue with MMP knockout animals [147]. Together with the recent finding that MT2-MMP integrates BM dynamics, integrin signaling, and growth factor activation to control epithelial proliferation and branching morphogenesis [51] demonstrates the critical functions of BM in SMG development.

Exogenous Matrices and Tissue Engineering

Tissue engineering research aimed at generating functional artificial salivary glands has investigated the use of various biocompatible and biodegradable polymers such as poly-(L)-lactic acid (PLLA), polyglycolic acid (PGA), poly-(L)-lactic-co-glycolic acid (PLGA) copolymer [148], polyvinyl alcohol (PVA), and chitosan [149–151], a deacetylated product of chitin that is structurally similar to GAGs, as cell substrates. These artificial substrates have been studied for their ability to support the differentiation and cellular function of embryonic salivary glands and salivary epithelial cell lines. Additionally, these polymers have been combined with purified ECM proteins, including FN, laminin, collagen I, collagen IV and gelatin, in an attempt to simulate the natural salivary gland microenvironment and to induce or maintain cell differentiation. Adult human salivary gland cell lines proliferate on PLLA coated with fibronectin or laminin but fail to form tight junctions even if transfected with tight junction proteins [148, 152, 153]. Stimulation of branching morphogenesis may be an important function of artificial matrices for some therapeutic strategies. Embryonic SMGs underwent a significant induction in branching morphogenesis when cultured on top of chitosan membranes [151, 154] and showed increased expression of collagen type III [151]. Stimulation of cellular secretion of ECM/BM may be important in engineering of a functioning salivary gland to generate a natural scaffold. Studies are underway to grow salivary gland cells on nanofibers, fibers made of artificial materials at the nanoscale to imitate the properties of ECM, to independently examine the role of surface topology, stiffness and chemical signals on salivary gland behavior [155].

Some tissue engineering approaches may require remodeling of the substrate, as occurs with ECM. Single embryonic SMG epithelial cells self-organized into a tissue aggregate that could continue to undergo branching morphogenesis and differentiation when grown in Matrigel [115]. Human SMG cells have a similar capacity when grown in Matrigel/collagen matrices [156]. These studies point to the ability of cells to self-organize when provided an appropriate matrix substrate. A key question facing tissue engineering is how to create an ideal cell substrate. That the ECM can be a critical component of the substrate is clear, but mesenchymal factors are also likely to be important. That mesenchymal-derived factors are important for cell differentiation was shown by the fact that mesenchymal cells promoted differentiation beyond Matrigel itself in self-organized SMG tissue aggregates [115]. The relationships between substrates and cell differentiation and morphogenesis are critical to our understanding of tissue assembly and tissue engineering and will be the topic of many future studies.

ECM Alterations in Sjögren's Syndrome

Sjögren's syndrome (SS) is a chronic systemic autoimmune disorder associated with extensive lymphocytic infiltration of both salivary and lacrimal tissues and loss of moisture production. That alterations in the BM are associated with pathogenesis of SS was not known until recently. SS is one of the three most common autoimmune disorders and affects predominantly women (prevalent in as many as 1 out of every 2,500 females), with a typical onset around 40–50 years of age [157]. SS may be classified as primary when it involves only decreased production of saliva and tears, or as secondary, when manifesting along with other autoimmune conditions, such as such as rheumatoid arthritis, systemic lupus erythematosus, scleroderma, polymyositis, and polyarteritis nodosa.

Details regarding the specific cellular and molecular alterations that occur in salivary gland cells in SS are currently lacking. However, recent studies using both human SS patient tissue and mouse models of SS have begun to elucidate the changes occurring in both acinar and ductal morphology and function that may relate to ECM remodeling during disease occurrence and progression. Multi-photon imaging with second harmonic generation to observe fibrils in thick tissue demonstrated increased degradation of ECM proteins including collagen, elastin and GAGs in lacrimal tissue [158]. It was observed that expression of both laminin and type IV collagen was reduced in the acinar and ductal cells of labial glands of SS patients [158]. Specifically, weaker type IV collagen α_5 and α_6 chain expression has been observed in SS glands [159]. In another study, severe disruption of integrin $\alpha_6\beta_4$ BM localization was observed in SS patients, although mRNA levels were unchanged [160]. Integrins α_1 and α_2 were found to be weakly expressed in acinar cells in SS but not in control salivary tissue [161]; the cellular consequences of this decreased expression are not yet understood.

Increasing evidence implicates proteolytic activity as a causative factor in SS. Overexpression of MMP-3 and MMP-9 is observed in SS [162]. Increased MMP expression has been suggested to contribute to the disorganization and disruption of BM continuity observed in SS acini and ducts, as well as to facilitate infiltration of cytotoxic T lymphocytes and hamper adhesion of acinar and ductal cells to ECM proteins [162, 163]. MMP activity is also known to be increased in salivary tissue in SS; particularly imbalanced ratios of MMP-9/TIMP-1 and MMP-3/TIMP-1 have been observed [164]. The proteolytic cleavage of decorin and biglycan was enhanced in both saliva and in glandular lysates in mouse models of SS. This was accompanied by elevated levels of TGF-β, for which decorin can also serve as a reservoir by sequestering it in the ECM away from its cell-surface receptor. This proteolytic activity was

MMP-driven since inhibition of MMP activity with pharmacological inhibitors could reduce TGF-β levels in a SS mouse model [165]. Further evidence of ECM remodeling involvement in SS was demonstrated by the finding that primary human salivary gland epithelial cells undergo anoikis when exposed to anti-Ro/SSA autoantibodies, concomitant with a reduction in fibulin-6 mRNA expression [166, 167]. MMP-associated degradation of BM proteins is a newly discovered change associated with the SS pathology that merits further investigation.

Older studies identified increased laminin expression by salivary ductal cells in SS patients [168]. Since then, additional studies have examined the expression of the laminin α_1, α_4, and γ_2 chains and nidogen mRNA and protein levels in both primary and secondary SS patients. Increases in the amount of mRNA and protein levels of both the processed and unprocessed laminin-γ_2 chain were increased in primary SS, while nidogen mRNA and protein levels were similar in SS patients and controls. Interestingly, high levels of nidogen degradation were observed in patients with low interacinar fibrosis while laminin α_1 and α_4 protein levels were increased in primary but not secondary SS patients, suggesting that active BM remodeling may occur during the earlier stages of SS disease progression [169]. Whether the deregulated expression and remodeling of these and other ECM proteins is a cause or an effect of the altered salivary gland morphology and function that is characteristic of SS, has yet to be confirmed.

ECM in Salivary Neoplasms

Most salivary gland neoplasms are benign and occur largely in the parotid gland, with the majority of those identified (80%) being unilateral pleomorphic adenomas with defined margins. However, malignant salivary gland tumor incidence increases with age and frequently occurs in the submandibular and sublingual glands. The most common malignant salivary gland tumors are mucoepidermoid carcinomas and adenoid cystic carcinomas followed by acinic cell carcinoma, adenocarcinoma, squamous cell carcinoma, and pleomorphic adenomas [170]. The close relationship between salivary glands and their associated ECM governs normal architecture and function of the glands, in addition to providing a physical scaffold for cell attachment. Aberrant production of ECM and BM proteins such as laminins, collagen type IV, FN, and tenascin have been detected in multiple salivary tumor types, and their expression levels are frequently augmented [171].

Adenoid cystic carcinomas (AdCCs), metastatic neoplasms that arise in secretory glands, have been shown to display increased ECM protein expression including collagen type IV, laminins, HSPGs, and entactin. Immunohistochemical staining techniques have been used to identify and classify these salivary gland tumors based on their characteristic ECM protein expression patterns in combination with cytological morphology [172, 173]. Excess ECM proteins can accumulate in the intercellular spaces, resulting in the formation of pseudocysts, a characteristic feature of AdCC

[174]. Perlecan is synthesized by proliferative AdCC cells, which have the ability to form stromal pseudocyst structures [175]. Further, AdCC-derived cell lines display enhanced collagen-stimulated migration, attributable in part to higher levels of urokinase-type plasminogen activator receptor (uPAR) and associated cell-surface signaling complexes [176]. The underlying cause or function of enhanced ECM synthesis in salivary tumors is not well understood, however the presence of excess ECM molecules could contribute to salivary tumor cell survival, proliferation, invasion, or metastasis. Interestingly, very little BM disruption, a hallmark of many metastatic tumors, was observed in malignant salivary tumor types [171]. The contribution of the ECM to salivary tumor initiation and progression merits further study.

Future Directions

We have made significant advances in our understanding of the basic mechanisms by which the ECM and BM influence the morphogenesis and differentiation of salivary glands; however, there is still much to be learned. Conditional knockout animals will continue to be useful to study the function of ECM molecules specifically during SMG development. Further profiling of both gene and protein expression will provide knowledge regarding temporal expression patterns during SMG development. In addition, continued expansion of molecular tools will aid in the study of ECM functions in ex vivo organ culture systems and elucidation of downstream signaling. It will be interesting to test what function the newly discovered short non-coding micro RNAs (miRNAs) [177] might have in regulation of ECM protein expression during development, and how deregulation of miRNAs might affect ECM proteins during disease pathogenesis.

Recent work has begun to examine salivary gland development at the systems level [178], although the function of the ECM has not been specifically considered in studies to date. Branching morphogenesis is a physical process, and therefore, can be described theoretically with mathematical modeling. In fact, studies have been initiated to develop theoretical mathematical models of branching morphogenesis [179, 180], although these studies have not specifically modeled properties of the ECM. Although studies of collagens and FN in SMG development have implied a function for molecular forces, such forces have not yet been measured. Measurement of morphogenetic forces using developing techniques, such as atomic force microscopy, will be required to develop realistic and accurate models. Mathematical models will ultimately aid in our systems level understanding of the process of branching morphogenesis.

Recent studies have indicated that ECM and BM proteins have clinical relevance for disease progression and therapeutic strategies. Given recent studies implicating ECM/BM remodeling in Sjögren's syndrome and salivary gland cancers, further investigation is required to understand whether such changes are a cause or an effect of disease progression, and if these changes can be used as diagnostic indicators. Development of a better understanding of ECM-mediated signaling will inform tissue engineering and regenerative strategies. The more closely cellular substrates can mimic the functions of ECM and BM in vivo, the more likely are the resulting engineered tissues to achieve secretory function. A greater understanding of the mechanisms by which ECM/BM proteins affect cellular organization and differentiation could lead to improvements in our understanding of tissue formation, tissue engineering, as well as disease initiation and progression.

Acknowledgements

The authors would like to thank Matthew Hoffman and Guy Carpenter for valuable comments and Anthony Hickey for assistance in preparing the figures. This work was partially supported by NIH grants DE019244401 and DE019197 (to M.L.) and by the University at Albany, State University of New York.

References

1 Patel VN, Rebustini IT, Hoffman MP: Salivary gland branching morphogenesis. Differentiation 2006;74:349–364.
2 Tucker AS: Salivary gland development. Semin Cell Dev Biol 2007;18:237–244.
3 Grobstein C: Inductive epitheliomesenchymal interaction in cultured organ rudiments of the mouse. Science 1953;118:52–55.
4 Borghese E: The development in vitro of the submandibular and sublingual glands of *Mus musculus*. J Anat 1950;84:287–302.
5 Cutler LS: The dependent and independent relationships between cytodifferentiation and morphogenesis in developing salivary gland secretory cells. Anat Rec 1980;196:341–347.
6 Daley WP, Peters SB, Larsen M: Extracellular matrix dynamics in development and regenerative medicine. J Cell Sci 2008;121:255–264.
7 Larsen M, Artym VV, Green JA, Yamada KM: The matrix reorganized: extracellular matrix remodeling and integrin signaling. Curr Opin Cell Biol 2006;18:463–471.
8 LeBleu VS, Macdonald B, Kalluri R: Structure and function of basement membranes. Exp Biol Med (Maywood) 2007;232:1121–1129.
9 Stanley JR, Woodley DT, Katz SI, Martin GR: Structure and function of basement membrane. J Invest Dermatol 1982;79(suppl 1):69S–72S.
10 Kadoya Y, Yamashina S: Ultrastructure of the basement membrane and its precursor in developing rat submandibular gland as shown by alcian blue staining. Cell Tissue Res 1992;268:233–238.
11 Bernfield MR, Banerjee SD: Acid mucopolysaccharide (glycosaminoglycan) at the epithelial-mesenchymal interface of mouse embryo salivary glands. J Cell Biol 1972;52:664–673.
12 Bernfield MR, Banerjee SD, Cohn RH: Dependence of salivary epithelial morphology and branching morphogenesis upon acid mucopolysaccharide-protein (proteoglycan) at the epithelial surface. J Cell Biol 1972;52:674–689.
13 Gullberg D, Ekblom P: Extracellular matrix and its receptors during development. Int J Dev Biol 1995;39:845–854.
14 Banerjee SD, Cohn RH, Bernfield MR: Basal lamina of embryonic salivary epithelia. Production by the epithelium and role in maintaining lobular morphology. J Cell Biol 1977;73:445–463.
15 Brownell AG, Slavkin HC: Role of basal lamina in tissue interactions. Ren Physiol 1980;3:193–204.
16 Segawa A, Sahara N, Suzuki K, Yamashina S: Acinar structure and membrane regionalization as a prerequisite for exocrine secretion in the rat submandibular gland. J Cell Sci 1985;78:67–85.
17 Bernfield M, Banerjee SD: The turnover of basal lamina glycosaminoglycan correlates with epithelial morphogenesis. Dev Biol 1982;90:291–305.
18 Bernfield M, Banerjee SD, Koda JE, Rapraeger AC: Remodelling of the basement membrane: morphogenesis and maturation. Ciba Found Symp 1984;108:179–196.
19 Larsen M, Wei C, Yamada KM: Cell and fibronectin dynamics during branching morphogenesis. J Cell Sci 2006;119:3376–3384.

20 Nakanishi Y, Ishii T: Epithelial shape change in mouse embryonic submandibular gland: modulation by extracellular matrix components. Bioessays 1989;11:163–167.

21 Grobstein C, Cohen J: Collagenase: effect on the morphogenesis of embryonic salivary epithelium in vitro. Science 1965;150:626–628.

22 Brownell AG, Bessem CC, Slavkin HC: Possible functions of mesenchyme cell-derived fibronectin during formation of basal lamina. Proc Natl Acad Sci USA 1981;78:3711–3715.

23 Kadoya Y, Yamashina S: Intracellular accumulation of basement membrane components during morphogenesis of rat submandibular gland. J Histochem Cytochem 1989;37:1387–1392.

24 Hardman P, Spooner BS: Localization of extracellular matrix components in developing mouse salivary glands by confocal microscopy. Anat Rec 1992; 234:452–459.

25 Kadoya Y, Salmivirta K, Talts JF, Kadoya K, Mayer U, Timpl R, Ekblom P: Importance of nidogen binding to laminin γ_1 for branching epithelial morphogenesis of the submandibular gland. Development 1997;124:683–691.

26 Ekblom P, Ekblom M, Fecker L, Klein G, Zhang HY, Kadoya Y, Chu ML, Mayer U, Timpl R: Role of mesenchymal nidogen for epithelial morphogenesis in vitro. Development 1994;120:2003–2014.

27 Costell M, Gustafsson E, Aszodi A, Morgelin M, Bloch W, Hunziker E, Addicks K, Timpl R, Fassler R: Perlecan maintains the integrity of cartilage and some basement membranes. J Cell Biol 1999;147: 1109–1122.

28 Arikawa-Hirasawa E, Watanabe H, Takami H, Hassell JR, Yamada Y: Perlecan is essential for cartilage and cephalic development. Nat Genet 1999;23:354–358.

29 Sakai T, Larsen M, Yamada KM: Fibronectin requirement in branching morphogenesis. Nature 2003;423:876–881.

30 Borghese E: Explantation experiments on the influence of the connective tissue capsule on the development of the epithelial part of the submandibular gland of *Mus musculus*. J Anat 1950;84:303–318.

31 Kusakabe M, Sakakura T, Sano M, Nishizuka Y: A pituitary-salivary mixed gland induced by tissue recombination of embryonic pituitary epithelium and embryonic submandibular gland mesenchyme in mice. Dev Biol 1985;110:382–391.

32 Grobstein C: Morphogenetic interaction between embryonic mouse tissues separated by a membrane filter. Nature 1953;172:869–870.

33 Takahashi Y, Nogawa H: Branching morphogenesis of mouse salivary epithelium in basement membrane-like substratum separated from mesenchyme by the membrane filter. Development 1991;111:327–335.

34 Nogawa H, Takahashi Y: Substitution for mesenchyme by basement-membrane-like substratum and epidermal growth factor in inducing branching morphogenesis of mouse salivary epithelium. Development 1991;112:855–861.

35 Rebustini IT, Hoffman MP: ECM and FGF-dependent assay of embryonic SMG epithelial morphogenesis: investigating growth factor/matrix regulation of gene expression during submandibular gland development. Methods Mol Biol 2009;522:319–330.

36 Kleinman HK, McGarvey ML, Hassell JR, Star VL, Cannon FB, Laurie GW, Martin GR: Basement membrane complexes with biological activity. Biochemistry 1986;25:312–318.

37 Benton G, George J, Kleinman HK, Arnaoutova IP: Advancing science and technology via 3D culture on basement membrane matrix. J Cell Physiol 2009; 221:18–25.

38 Kleinman HK, Martin GR: Matrigel: basement membrane matrix with biological activity. Semin Cancer Biol 2005;15:378–386.

39 Hardman P, Klement BJ, Spooner BS: Growth and morphogenesis of embryonic mouse organs on non-coated and extracellular matrix-coated Biopore membrane. Dev Growth Differ 1993;35:683–690.

40 Royce LS, Kibbey MC, Mertz P, Kleinman HK, Baum BJ: Human neoplastic submandibular intercalated duct cells express an acinar phenotype when cultured on a basement membrane matrix. Differentiation 1993;52:247–255.

41 Zheng C, Hoffman MP, McMillan T, Kleinman HK, O'Connell BC: Growth factor regulation of the amylase promoter in a differentiating salivary acinar cell line. J Cell Physiol 1998;177:628–635.

42 Nakanishi Y, Sugiura F, Kishi J, Hayakawa T: Collagenase inhibitor stimulates cleft formation during early morphogenesis of mouse salivary gland. Dev Biol 1986;113:201–206.

43 Bernfield MR, Wessells NK: Intra- and extracellular control of epithelial morphogenesis. Symp Soc Dev Biol 1970;29:195–249.

44 Spooner BS, Faubion JM: Collagen involvement in branching morphogenesis of embryonic lung and salivary gland. Dev Biol 1980;77:84–102.

45 Nakanishi Y, Sugiura F, Kishi J, Hayakawa T: Scanning electron microscopic observation of mouse embryonic submandibular glands during initial branching: preferential localization of fibrillar structures at the mesenchymal ridges participating in cleft formation. J Embryol Exp Morphol 1986;96:65–77.

46 Fukuda Y, Masuda Y, Kishi J, Hashimoto Y, Hayakawa T, Nogawa H, Nakanishi Y: The role of interstitial collagens in cleft formation of mouse embryonic submandibular gland during initial branching. Development 1988;103:259–267.

47 Nakanishi Y, Nogawa H, Hashimoto Y, Kishi J, Hayakawa T: Accumulation of collagen III at the cleft points of developing mouse submandibular epithelium. Development 1988;104:51–59.

48 Kratochwil K, Dziadek M, Lohler J, Harbers K, Jaenisch R: Normal epithelial branching morphogenesis in the absence of collagen I. Dev Biol 1986;117:596–606.

49 Liu X, Wu H, Byrne M, Krane S, Jaenisch R: Type III collagen is crucial for collagen I fibrillogenesis and for normal cardiovascular development. Proc Natl Acad Sci USA 1997;94:1852–1856.

50 Cutler LS: The role of extracellular matrix in the morphogenesis and differentiation of salivary glands. Adv Dent Res 1990;4:27–33.

51 Rebustini IT, Myers C, Lassiter KS, Surmak A, Szabova L, Holmbeck K, Pedchenko V, Hudson BG, Hoffman MP: MT2-MMP-dependent release of collagen IV NC1 domains regulates submandibular gland branching morphogenesis. Dev Cell 2009;17: 482–493.

52 Miner JH: Laminins and their roles in mammals. Microsc Res Tech 2008;71:349–356.

53 Tzu J, Marinkovich MP: Bridging structure with function: structural, regulatory, and developmental role of laminins. Int J Biochem Cell Biol 2008;40: 199–214.

54 Aumailley M, Bruckner-Tuderman L, Carter WG, Deutzmann R, Edgar D, Ekblom P, Engel J, Engvall E, Hohenester E, Jones JC, Kleinman HK, Marinkovich MP, Martin GR, Mayer U, Meneguzzi G, Miner JH, Miyazaki K, Patarroyo M, Paulsson M, Quaranta V, Sanes JR, Sasaki T, Sekiguchi K, Sorokin LM, Talts JF, Tryggvason K, Uitto J, Virtanen I, von der Mark K, Wewer UM, Yamada Y, Yurchenco PD: A simplified laminin nomenclature. Matrix Biol 2005;24:326–332.

55 Sonnenberg A, Linders CJ, Modderman PW, Damsky CH, Aumailley M, Timpl R: Integrin recognition of different cell-binding fragments of laminin (P1, E3, E8) and evidence that $\alpha_6\beta_1$ but not $\alpha_6\beta_4$ functions as a major receptor for fragment E8. J Cell Biol 1990;110:2145–2155.

56 Suzuki N, Ichikawa N, Kasai S, Yamada M, Nishi N, Morioka H, Yamashita H, Kitagawa Y, Utani A, Hoffman MP, Nomizu M: Syndecan binding sites in the laminin α_1 chain G domain. Biochemistry 2003; 42:12625–12633.

57 Durbeej M, Talts JF, Henry MD, Yurchenco PD, Campbell KP, Ekblom P: Dystroglycan binding to laminin α_1 LG4 module influences epithelial morphogenesis of salivary gland and lung in vitro. Differentiation 2001;69:121–134.

58 Kadoya Y, Yamashina S: Salivary gland morphogenesis and basement membranes. Anat Sci Int 2005;80: 71–79.

59 Kadoya Y, Kadoya K, Durbeej M, Holmvall K, Sorokin L, Ekblom P: Antibodies against domain E3 of laminin-1 and integrin α_6 subunit perturb branching epithelial morphogenesis of submandibular gland, but by different modes. J Cell Biol 1995; 129:521–534.

60 Hosokawa Y, Takahashi Y, Kadoya Y, Yamashina S, Nomizu M, Yamada Y, Nogawa H: Significant role of laminin-1 in branching morphogenesis of mouse salivary epithelium cultured in basement membrane matrix. Dev Growth Differ 1999;41:207–216.

61 Kadoya Y, Nomizu M, Sorokin LM, Yamashina S, Yamada Y: Laminin α_1 chain G domain peptide, RKRLQVQLSIRT, inhibits epithelial branching morphogenesis of cultured embryonic mouse submandibular gland. Dev Dyn 1998;212:394–402.

62 Miner JH, Li C, Mudd JL, Go G, Sutherland AE: Compositional and structural requirements for laminin and basement membranes during mouse embryo implantation and gastrulation. Development 2004;131:2247–2256.

63 Rebustini IT, Patel VN, Stewart JS, Layvey A, Georges-Labouesse E, Miner JH, Hoffman MP: Laminin α5 is necessary for submandibular gland epithelial morphogenesis and influences FGFR expression through β_1 integrin signaling. Dev Biol 2007;308:15–29.

64 Kadoya Y, Mochizuki M, Nomizu M, Sorokin L, Yamashina S: Role for laminin-α_5 chain LG4 module in epithelial branching morphogenesis. Dev Biol 2003;263:153–164.

65 Miner JH, Cunningham J, Sanes JR: Roles for laminin in embryogenesis: exencephaly, syndactyly, and placentopathy in mice lacking the laminin α5 chain. J Cell Biol 1998;143:1713–1723.

66 Ho MS, Bose K, Mokkapati S, Nischt R, Smyth N: Nidogens – extracellular matrix linker molecules. Microsc Res Tech 2008;71:387–395.

67 Ries A, Gohring W, Fox JW, Timpl R, Sasaki T: Recombinant domains of mouse nidogen-1 and their binding to basement membrane proteins and monoclonal antibodies. Eur J Biochem 2001;268: 5119–5128.

68 De Vega S, Iwamoto T, Yamada Y: Fibulins: multiple roles in matrix structures and tissue functions. Cell Mol Life Sci 2009;66:1890–1902.

69 Dziadek M, Timpl R: Expression of nidogen and laminin in basement membranes during mouse embryogenesis and in teratocarcinoma cells. Dev Biol 1985;111:372–382.

70 Miosge N, Quondamatteo F, Klenczar C, Herken R: Nidogen-1: expression and ultrastructural localization during the onset of mesoderm formation in the early mouse embryo. J Histochem Cytochem 2000; 48:229–238.

71 Murshed M, Smyth N, Miosge N, Karolat J, Krieg T, Paulsson M, Nischt R: The absence of nidogen-1 does not affect murine basement membrane formation. Mol Cell Biol 2000;20:7007–7012.

72 Schymeinsky J, Nedbal S, Miosge N, Poschl E, Rao C, Beier DR, Skarnes WC, Timpl R, Bader BL: Gene structure and functional analysis of the mouse nidogen-2 gene: nidogen-2 is not essential for basement membrane formation in mice. Mol Cell Biol 2002; 22:6820–6830.

73 Bader BL, Smyth N, Nedbal S, Miosge N, Baranowsky A, Mokkapati S, Murshed M, Nischt R: Compound genetic ablation of nidogen 1 and 2 causes basement membrane defects and perinatal lethality in mice. Mol Cell Biol 2005;25:6846–6856.

74 Steinberg Z, Myers C, Heim VM, Lathrop CA, Rebustini IT, Stewart JS, Larsen M, Hoffman MP: FGFR2b signaling regulates ex vivo submandibular gland epithelial cell proliferation and branching morphogenesis. Development 2005;132:1223–1234.

75 Battaglia C, Mayer U, Aumailley M, Timpl R: Basement-membrane heparan sulfate proteoglycan binds to laminin by its heparan sulfate chains and to nidogen by sites in the protein core. Eur J Biochem 1992;208:359–366.

76 Kohfeldt E, Sasaki T, Gohring W, Timpl R: Nidogen-2: a new basement membrane protein with diverse binding properties. J Mol Biol 1998;282:99–109.

77 Hirai Y: Epimorphin as a morphogen: does a protein for intracellular vesicular targeting act as an extracellular signaling molecule? Cell Biol Int 2001; 25:193–195.

78 Hirai Y, Takebe K, Takashina M, Kobayashi S, Takeichi M: Epimorphin: a mesenchymal protein essential for epithelial morphogenesis. Cell 1992;69: 471–481.

79 Wang Y, Wang L, Iordanov H, Swietlicki EA, Zheng Q, Jiang S, Tang Y, Levin MS, Rubin DC: Epimorphin(–/–) mice have increased intestinal growth, decreased susceptibility to dextran sodium sulfate colitis, and impaired spermatogenesis. J Clin Invest 2006;116:1535–1546.

80 Bornstein P, Sage EH: Matricellular proteins: extracellular modulators of cell function. Curr Opin Cell Biol 2002;14:608–616.

81 Ito Y, Tamura I: Age-related changes in collagen, laminin and tenascin in the infant rat submandibular gland. J Osaka Dent Univ 1997;31:19–27.

82 Saga Y, Yagi T, Ikawa Y, Sakakura T, Aizawa S: Mice develop normally without tenascin. Genes Dev 1992;6:1821–1831.

83 Furuse C, Cury PR, de Araujo NS, de Araujo VC: Immunoexpression of extracellular matrix proteins in human salivary gland development. Eur J Oral Sci 2004;112:548–551.

84 Thompson HA, Spooner BS: Inhibition of branching morphogenesis and alteration of glycosaminoglycan biosynthesis in salivary glands treated with β-D-xyloside. Dev Biol 1982;89:417–424.

85 Spooner BS, Bassett K, Stokes B: Sulfated glycosaminoglycan deposition and processing at the basal epithelial surface in branching and β-D-xyloside-inhibited embryonic salivary glands. Dev Biol 1985;109:177–183.

86 Spooner BS, Thompson-Pletscher HA, Stokes B, Bassett KE: Extracellular matrix involvement in epithelial branching morphogenesis. Dev Biol (NY 1985) 1986;3:225–260.

87 Ruoslahti E, Yamaguchi Y: Proteoglycans as modulators of growth factor activities. Cell 1991;64:867–869.

88 Patel VN, Knox SM, Likar KM, Lathrop CA, Hossain R, Eftekhari S, Whitelock JM, Elkin M, Vlodavsky I, Hoffman MP: Heparanase cleavage of perlecan heparan sulfate modulates FGF10 activity during ex vivo submandibular gland branching morphogenesis. Development 2007;134:4177–4186.

89 Patel VN, Likar KM, Zisman-Rozen S, Cowherd SN, Lassiter KS, Sher I, Yates EA, Turnbull JE, Ron D, Hoffman MP: Specific heparan sulfate structures modulate FGF10-mediated submandibular gland epithelial morphogenesis and differentiation. J Biol Chem 2008;283:9308–9317.

90 Umeda Y, Miyazaki Y, Shiinoki H, Higashiyama S, Nakanishi Y, Hieda Y: Involvement of heparin-binding EGF-like growth factor and its processing by metalloproteinases in early epithelial morphogenesis of the submandibular gland. Dev Biol 2001; 237:202–211.

91 Hopf M, Gohring W, Kohfeldt E, Yamada Y, Timpl R: Recombinant domain IV of perlecan binds to nidogens, laminin-nidogen complex, fibronectin, fibulin-2 and heparin. Eur J Biochem 1999;259:917–925.

92 Knox SM, Whitelock JM: Perlecan: how does one molecule do so many things? Cell Mol Life Sci 2006; 63:2435–2445.

93 Muragaki Y, Timmons S, Griffith CM, Oh SP, Fadel B, Quertermous T, Olsen BR: Mouse Col18a1 is expressed in a tissue-specific manner as three alternative variants and is localized in basement membrane zones. Proc Natl Acad Sci USA 1995;92: 8763–8767.

94 Halfter W, Dong S, Schurer B, Cole GJ: Collagen XVIII is a basement membrane heparan sulfate proteoglycan. J Biol Chem 1998;273:25404–25412.

95 Saksela O, Moscatelli D, Sommer A, Rifkin DB: Endothelial cell-derived heparan sulfate binds basic fibroblast growth factor and protects it from proteolytic degradation. J Cell Biol 1988;107:743–751.

96 Pradhan S, Zhang C, Jia X, Carson D, Witt RL, Farach-Carson MC: Perlecan domain IV peptide stimulates salivary gland cell assembly in vitro. Tissue Eng Part A 2009;15:3309–3320.
97 Cohn RH, Banerjee SD, Bernfield MR: Basal lamina of embryonic salivary epithelia. Nature of glycosaminoglycan and organization of extracellular materials. J Cell Biol 1977;73:464–478.
98 Cutler LS, Christian CP, Rendell JK: Glycosaminoglycan synthesis by adult rat submandibular salivary-gland secretory units. Arch Oral Biol 1987;32:413–419.
99 Thompson HA, Spooner BS: Proteoglycan and glycosaminoglycan synthesis in embryonic mouse salivary glands: effects of β-D-xyloside, an inhibitor of branching morphogenesis. J Cell Biol 1983;96:1443–1450.
100 Yasuko M, Keiichi Y, Toshiteru M, Yasuo N: Branching morphogenesis of mouse embryonic submandibular epithelia cultured under three different conditions. Dev Growth Diff 1994;36:529–539.
101 Lafrenie RM, Bernier SM, Yamada KM: Adhesion to fibronectin or collagen I gel induces rapid, extensive, biosynthetic alterations in epithelial cells. J Cell Physiol 1998;175:163–173.
102 Mao Y, Schwarzbauer JE: Fibronectin fibrillogenesis, a cell-mediated matrix assembly process. Matrix Biol 2005;24:389–399.
103 Pankov R, Yamada KM: Fibronectin at a glance. J Cell Sci 2002;115:3861–3863.
104 Pytela R, Pierschbacher MD, Ruoslahti E: Identification and isolation of a 140 kDa cell surface glycoprotein with properties expected of a fibronectin receptor. Cell 1985;40:191–198.
105 Bernfield M, Sanderson RD: Syndecan, a developmentally regulated cell surface proteoglycan that binds extracellular matrix and growth factors. Philos Trans R Soc Lond Biol Soc 1990;327:171–186.
106 George EL, Georges-Labouesse EN, Patel-King RS, Rayburn H, Hynes RO: Defects in mesoderm, neural tube and vascular development in mouse embryos lacking fibronectin. Development 1993;119:1079–1091.
107 Daley WP, Gulfo KM, Sequeira SJ, Larsen M: Identification of a mechanochemical checkpoint and negative feedback loop regulating branching morphogenesis. Dev Biol 2009;336:169–182.
108 Lafrenie RM, Yamada KM: Integrins and matrix molecules in salivary gland cell adhesion, signaling, and gene expression. Ann NY Acad Sci 1998;842:42–48.
109 Bradley PL, Myat MM, Comeaux CA, Andrew DJ: Posterior migration of the salivary gland requires an intact visceral mesoderm and integrin function. Dev Biol 2003;257:249–262.
110 Kadoya Y, Yamashina S: Distribution of α_6 integrin subunit in developing mouse submandibular gland. J Histochem Cytochem 1993;41:1707–1714.
111 Franchi A, Santoro R, Paglierani M, Bondi R: Immunolocalization of α_2, α_5, and α_6 integrin subunits in salivary tissue and adenomas of the parotid gland. J Oral Pathol Med 1994;23:457–460.
112 Lazowski KW, Mertz PM, Redman RS, Kousvelari E: Temporal and spatial expression of laminin, collagen types IV and I and α_6/β_1 integrin receptor in the developing rat parotid gland. Differentiation 1994;56:75–82.
113 Sunardhi-Widyaputra S, Van Damme B: Distribution of the VLA family of integrins in normal salivary gland and in pleomorphic adenoma. Pathol Res Pract 1994;190:600–608.
114 Lourenco SV, Lima DM, Uyekita SH, Schultz R, de Brito T: Expression of β_1 integrin in human developing salivary glands and its parallel relation with maturation markers: in situ hybridisation and immunofluorescence study. Arch Oral Biol 2007;52:1064–1071.
115 Wei C, Larsen M, Hoffman MP, Yamada KM: Self-organization and branching morphogenesis of primary salivary epithelial cells. Tissue Eng 2007;13:721–735.
116 Menko AS, Kreidberg JA, Ryan TT, Van Bockstaele E, Kukuruzinska MA: Loss of $\alpha_3\beta_1$ integrin function results in an altered differentiation program in the mouse submandibular gland. Dev Dyn 2001;220:337–349.
117 Ekblom P: Receptors for laminins during epithelial morphogenesis. Curr Opin Cell Biol 1996;8:700–706.
118 Durbeej M, Larsson E, Ibraghimov-Beskrovnaya O, Roberds SL, Campbell KP, Ekblom P: Non-muscle α-dystroglycan is involved in epithelial development. J Cell Biol 1995;130:79–91.
119 Morais Freitas V, Nogueira da Gama de Souza L, Cyreno Oliveira E, Furuse C, Cavalcanti de Araujo V, Gastaldoni Jaeger R: Malignancy-related 67 kDa laminin receptor in adenoid cystic carcinoma. Effect on migration and β-catenin expression. Oral Oncol 2007;43:987–998.
120 Bernfield M, Kokenyesi R, Kato M, Hinkes MT, Spring J, Gallo RL, Lose EJ: Biology of the syndecans: a family of transmembrane heparan sulfate proteoglycans. Annu Rev Cell Biol 1992;8:365–393.
121 Stepp MA, Gibson HE, Gala PH, Iglesia DD, Pajoohesh-Ganji A, Pal-Ghosh S, Brown M, Aquino C, Schwartz AM, Goldberger O, Hinkes MT, Bernfield M: Defects in keratinocyte activation during wound healing in the syndecan-1-deficient mouse. J Cell Sci 2002;115:4517–4531.

122 Stepp MA, Liu Y, Pal-Ghosh S, Jurjus RA, Tadvalkar G, Sekaran A, Losicco K, Jiang L, Larsen M, Li L, Yuspa SH: Reduced migration, altered matrix and enhanced TGF-β_1 signaling are signatures of mouse keratinocytes lacking Sdc1. J Cell Sci 2007;120:2851–2863.

123 Hoffman MP, Nomizu M, Roque E, Lee S, Jung DW, Yamada Y, Kleinman HK: Laminin-1 and laminin-2 G-domain synthetic peptides bind syndecan-1 and are involved in acinar formation of a human submandibular gland cell line. J Biol Chem 1998;273: 28633–28641.

124 Makarenkova HP, Hoffman MP, Beenken A, Eliseenkova AV, Meech R, Tsau C, Patel VN, Lang RA, Mohammadi M: Differential interactions of FGFs with heparan sulfate control gradient formation and branching morphogenesis. Sci Signal 2009; 2:ra55.

125 Nie X, Luukko K, Kettunen P: FGF signalling in craniofacial development and developmental disorders. Oral Dis 2006;12:102–111.

126 Thisse B, Thisse C: Functions and regulations of fibroblast growth factor signaling during embryonic development. Dev Biol 2005;287:390–402.

127 Vasiliauskas D, Stern CD: Patterning the embryonic axis: FGF signaling and how vertebrate embryos measure time. Cell 2001;106:133–136.

128 Min H, Danilenko DM, Scully SA, Bolon B, Ring BD, Tarpley JE, DeRose M, Simonet WS: FGF-10 is required for both limb and lung development and exhibits striking functional similarity to *Drosophila* branchless. Genes Dev 1998;12:3156–3161.

129 Ohuchi H, Hori Y, Yamasaki M, Harada H, Sekine K, Kato S, Itoh N: FGF10 acts as a major ligand for FGF receptor 2 IIIb in mouse multi-organ development. Biochem Biophys Res Commun 2000;277:643–649.

130 De Moerlooze L, Spencer-Dene B, Revest JM, Hajihosseini M, Rosewell I, Dickson C: An important role for the IIIb isoform of fibroblast growth factor receptor 2 (FGFR2) in mesenchymal-epithelial signalling during mouse organogenesis. Development 2000;127:483–492.

131 Govindarajan V, Ito M, Makarenkova HP, Lang RA, Overbeek PA: Endogenous and ectopic gland induction by FGF-10. Dev Biol 2000;225:188–200.

132 Mailleux AA, Spencer-Dene B, Dillon C, Ndiaye D, Savona-Baron C, Itoh N, Kato S, Dickson C, Thiery JP, Bellusci S: Role of FGF10/FGFR2b signaling during mammary gland development in the mouse embryo. Development 2002;129:53–60.

133 Jaskoll T, Abichaker G, Witcher D, Sala FG, Bellusci S, Hajihosseini MK, Melnick M: FGF10/FGFR2b signaling plays essential roles during in vivo embryonic submandibular salivary gland morphogenesis. BMC Dev Biol 2005;5:11.

134 Makarenkova HP, Hoffman MP, Beenken A, Eliseenkova AV, Tsau C, Patel VN, Lang RA, Mohammadi M: Differential interactions of FGFs with heparan sulfate control gradient formation and branching morphogenesis. Sci Signal 2009;2:ra55.

135 Gresik EW, Kashimata M, Kadoya Y, Mathews R, Minami N, Yamashina S: Expression of epidermal growth factor receptor in fetal mouse submandibular gland detected by a biotinyltyramide-based catalyzed signal amplification method. J Histochem Cytochem 1997;45:1651–1657.

136 Kashimata M, Gresik EW: Epidermal growth factor system is a physiological regulator of development of the mouse fetal submandibular gland and regulates expression of the α6-integrin subunit. Dev Dyn 1997;208:149–161.

137 Forsten-Williams K, Chu CL, Fannon M, Buczek-Thomas JA, Nugent MA: Control of growth factor networks by heparan sulfate proteoglycans. Ann Biomed Eng 2008;36:2134–2148.

138 Werner H, Katz J: The emerging role of the insulin-like growth factors in oral biology. J Dent Res 2004; 83:832–836.

139 Katz J, Nasatzky E, Werner H, Le Roith D, Shemer J: Tumor necrosis factor α and interferon-γ-induced cell growth arrest is mediated via insulin-like growth factor binding protein-3. Growth Horm IGF Res 1999;9:174–178.

140 Conover CA, Khosla S: Role of extracellular matrix in insulin-like growth factor (IGF) binding protein-2 regulation of IGF-II action in normal human osteoblasts. Growth Horm IGF Res 2003;13:328–335.

141 Hansson HA, Nilsson A, Isgaard J, Billig H, Isaksson O, Skottner A, Andersson IK, Rozell B: Immunohistochemical localization of insulin-like growth factor i in the adult rat. Histochemistry 1988;89:403–410.

142 Amano O, Iseki S: Expression, localization and developmental regulation of insulin-like growth factor I MRNA in rat submandibular gland. Arch Oral Biol 1993;38:671–677.

143 Jaskoll T, Melnick M: Submandibular gland morphogenesis: Stage-specific expression of TGF-α/EGF, IGF, TGF-β, TNF, and IL-6 signal transduction in normal embryonic mice and the phenotypic effects of TGF-β_2, TGF-β_3, and EGF-R null mutations. Anat Rec 1999;256:252–268.

144 Melnick M, Jaskoll T: Mouse submandibular gland morphogenesis: a paradigm for embryonic signal processing. Crit Rev Oral Biol Med 2000;11:199–215.

145 Hayakawa T, Kishi J, Nakanishi Y: Salivary gland morphogenesis: possible involvement of collagenase. Matrix Suppl 1992;1:344–351.

146 Oblander SA, Zhou Z, Galvez BG, Starcher B, Shannon JM, Durbeej M, Arroyo AG, Tryggvason K, Apte SS: Distinctive functions of membrane type 1 matrix-metalloprotease (MT1-MMP or MMP-14) in lung and submandibular gland development are independent of its role in pro-MMP-2 activation. Dev Biol 2005;277:255–269.
147 Page-McCaw A, Ewald AJ, Werb Z: Matrix metalloproteinases and the regulation of tissue remodelling. Nat Rev 2007;8:221–233.
148 Aframian DJ, Cukierman E, Nikolovski J, Mooney DJ, Yamada KM, Baum BJ: The growth and morphological behavior of salivary epithelial cells on matrix protein-coated biodegradable substrata. Tissue Eng 2000;6:209–216.
149 Chen MH, Chen YJ, Liao CC, Chan YH, Lin CY, Chen RS, Young TH: Formation of salivary acinar cell spheroids in vitro above a polyvinyl alcohol-coated surface. J Biomed Mater Res A 2009;90:1066–1072.
150 Chen MH, Hsu YH, Lin CP, Chen YJ, Young TH: Interactions of acinar cells on biomaterials with various surface properties. J Biomed Mater Res A 2005;74:254–262.
151 Yang TL, Young TH: The enhancement of submandibular gland branch formation on chitosan membranes. Biomaterials 2008;29:2501–2508.
152 Aframian DJ, Palmon A: Current status of the development of an artificial salivary gland. Tissue Eng Part B Rev 2008;14:187–198.
153 Aframian DJ, Tran SD, Cukierman E, Yamada KM, Baum BJ: Absence of tight junction formation in an allogeneic graft cell line used for developing an engineered artificial salivary gland. Tissue Eng 2002;8:871–878.
154 Yang TL, Young TH: Chitosan cooperates with mesenchyme-derived factors in regulating salivary gland epithelial morphogenesis. J Cell Mol Med 2008.
155 Larsen M, Jean-Gilles R, Soscia D, Sequeira SJ, Melfi M, Gadre A, Castracane J: Development of nanofiber scaffolds for engineering an artificial salivary gland. Proc ASME First Global Conference on NanoEngineering for Medicine and Biology, in press, 2010.
156 Joraku A, Sullivan CA, Yoo J, Atala A: In-vitro reconstitution of three-dimensional human salivary gland tissue structures. Differentiation 2007;75:318–324.
157 Fox RI, Stern M, Michelson P: Update in Sjögren syndrome. Curr Opin Rheumatol 2000;12:391–398.
158 Schenke-Layland K, Xie J, Angelis E, Starcher B, Wu K, Riemann I, MacLellan WR, Hamm-Alvarez SF: Increased degradation of extracellular matrix structures of lacrimal glands implicated in the pathogenesis of Sjögren's syndrome. Matrix Biol 2008;27:53–66.
159 Poduval P, Sillat T, Virtanen I, Porola P, Konttinen YT: Abnormal basement membrane type IV collagen α-chain composition in labial salivary glands in Sjögren's syndrome. Arthritis Rheum 2009;60:938–945.
160 Velozo J, Aguilera S, Alliende C, Ewert P, Molina C, Perez P, Leyton L, Quest A, Brito M, Gonzalez S, Leyton C, Hermoso M, Romo R, Gonzalez MJ: Severe alterations in expression and localisation of $\alpha_6\beta_4$ integrin in salivary gland acini from patients with Sjögren syndrome. Ann Rheum Dis 2009;68:991–996.
161 Laine M, Virtanen I, Porola P, Rotar Z, Rozman B, Poduval P, Konttinen YT: Acinar epithelial cell laminin-receptors in labial salivary glands in Sjögren's syndrome. Clin Exp Rheumatol 2008;26:807–813.
162 Goicovich E, Molina C, Perez P, Aguilera S, Fernandez J, Olea N, Alliende C, Leyton C, Romo R, Leyton L, Gonzalez MJ: Enhanced degradation of proteins of the basal lamina and stroma by matrix metalloproteinases from the salivary glands of Sjögren's syndrome patients: correlation with reduced structural integrity of acini and ducts. Arthritis Rheum 2003;48:2573–2584.
163 Molina C, Alliende C, Aguilera S, Kwon YJ, Leyton L, Martinez B, Leyton C, Perez P, Gonzalez MJ: Basal lamina disorganisation of the acini and ducts of labial salivary glands from patients with Sjögren's syndrome: association with mononuclear cell infiltration. Ann Rheum Dis 2006;65:178–183.
164 Perez P, Kwon YJ, Alliende C, Leyton L, Aguilera S, Molina C, Labra C, Julio M, Leyton C, Gonzalez MJ: Increased acinar damage of salivary glands of patients with Sjögren's syndrome is paralleled by simultaneous imbalance of matrix metalloproteinase-3/tissue inhibitor of metalloproteinases-1 and matrix metalloproteinase-9/tissue inhibitor of metalloproteinases-1 ratios. Arthritis Rheum 2005;52:2751–2760.
165 Yamachika S, Brayer J, Oxford GE, Peck AB, Humphreys-Beher MG: Aberrant proteolytic digestion of biglycan and decorin by saliva and exocrine gland lysates from the NOD mouse model for autoimmune exocrinopathy. Clin Exp Rheumatol 2000;18:233–240.
166 Lisi S, D'Amore M, Scagliusi P, Mitolo V, Sisto M: Anti-Ro/SSA autoantibody-mediated regulation of extracellular matrix fibulins in human epithelial cells of the salivary gland. Scand J Rheumatol 2009;38:198–206.
167 Sisto M, D'Amore M, Lofrumento DD, Scagliusi P, D'Amore S, Mitolo V, Lisi S: Fibulin-6 expression and anoikis in human salivary gland epithelial cells: implications in Sjögren's syndrome. Int Immunol 2009;21:303–311.

168 McArthur CP, Fox NW, Kragel P: Monoclonal antibody detection of laminin in minor salivary glands of patients with Sjögren's syndrome. J Autoimmun 1993;6:649–661.
169 Kwon YJ, Perez P, Aguilera S, Molina C, Leyton L, Alliende C, Leyton C, Brito M, Romo R, Gonzalez MJ: Involvement of specific laminins and nidogens in the active remodeling of the basal lamina of labial salivary glands from patients with Sjögren's syndrome. Arthritis Rheum 2006;54:3465–3475.
170 Stenner M, Klussmann JP: Current update on established and novel biomarkers in salivary gland carcinoma pathology and the molecular pathways involved. Eur Arch Otorhinolaryngol 2009;266:333–341.
171 Raitz R, Martins MD, Araujo VC: A study of the extracellular matrix in salivary gland tumors. J Oral Pathol Med 2003;32:290–296.
172 Kawahara A, Harada H, Kage M, Yokoyama T, Kojiro M: Extracellular material in adenoid cystic carcinoma of the salivary glands: a comparative cytological study with other salivary myoepithelial tumors. Diagn Cytopathol 2004;31:14–18.
173 Kawahara A, Harada H, Kage M, Yokoyama T, Kojiro M: Characterization of the epithelial components in pleomorphic adenoma of the salivary gland. Acta Cytol 2002;46:1095–1100.
174 Shirasuna K, Saka M, Hayashido Y, Yoshioka H, Sugiura T, Matsuya T: Extracellular matrix production and degradation by adenoid cystic carcinoma cells: Participation of plasminogen activator and its inhibitor in matrix degradation. Cancer Res 1993; 53:147–152.
175 Kimura S, Cheng J, Ida H, Hao N, Fujimori Y, Saku T: Perlecan (heparan sulfate proteoglycan) gene expression reflected in the characteristic histological architecture of salivary adenoid cystic carcinoma. Virchows Arch 2000;437:122–128.
176 Abu-Ali S, Sugiura T, Takahashi M, Shiratsuchi T, Ikari T, Seki K, Hiraki A, Matsuki R, Shirasuna K: Expression of the urokinase receptor regulates focal adhesion assembly and cell migration in adenoid cystic carcinoma cells. J Cell Physiol 2005;203:410–419.
177 Jevnaker AM, Osmundsen H: MicroRNA expression profiling of the developing murine molar tooth germ and the developing murine submandibular salivary gland. Arch Oral Biol 2008;53:629–645.
178 Melnick M, Phair RD, Lapidot SA, Jaskoll T: Salivary gland branching morphogenesis: a quantitative systems analysis of the Eda/Edar/NFκB paradigm. BMC developmental biology 2009;9:32.
179 Lubkin SR: Branched organs: mechanics of morphogenesis by multiple mechanisms. Curr Top Dev Biol 2008;81:249–268.
180 Wan X, Li Z, Lubkin SR: Mechanics of mesenchymal contribution to clefting force in branching morphogenesis. Biomech Model Mechanobiol 2008; 7:417–426.

Melinda Larsen, PhD, Assist. Prof.
Department of Biological Sciences, University at Albany, SUNY
1400 Washington Ave, LSRB 1086, Albany, NY 12222 (USA)
Tel. +1 518 591 8882, Fax +1 518 442 4767
E-Mail mlarsen@albany.edu

Tucker AS, Miletich I (eds): Salivary Glands. Development, Adaptations and Disease.
Front Oral Biol. Basel, Karger, 2010, vol 14, pp 78–89

Lumen Formation in Salivary Gland Development

Kirsty L. Wells · Nisha Patel

Department of Craniofacial Development, Guy's Campus, King's College London, London, UK

Abstract

During salivary gland morphogenesis, the developing ducts and acini must hollow out to form lumina which will eventually allow the free passage and modification of saliva on its journey from acini to oral cavity. The molecular mechanisms that participate in the creation of this tubular structure are of great research interest. Histological studies show that lumen formation begins during the mid stages of branching morphogenesis. At this stage, apoptotic cells are detectable in the developing salivary ducts at sites where lumina are forming, suggesting that programmed cell death is instrumental in clearing the luminal space. The formation of cell-cell junctions between the epithelial cells lining the space is also an integral part of lumen formation, since these junctions form a barrier around the lumen and allow the surfaces of the lumen-lining cells to become specialized. This chapter will discuss the mechanisms involved in salivary gland lumen formation during development, and draw on the most recent research in this interesting field.

An important and often overlooked element of salivary gland morphogenesis is the creation of the luminal space. The interesting question of how a network of hollow epithelial tubes is formed during development is fundamental to understanding the morphogenesis of a wide variety of organs, and has been studied in such diverse structures as the *Drosophila* trachea and the murine mammary gland. In the salivary glands, a valuable model for the formation of lumina, initially solid epithelial branches hollow out to form ducts and acini, eventually allowing the free passage and modification of saliva on its journey from acini to oral cavity.

The Histology of Lumen Formation

Lumen formation begins during embryonic development of the salivary gland, taking place alongside branching morphogenesis. The temporal progression of lumen

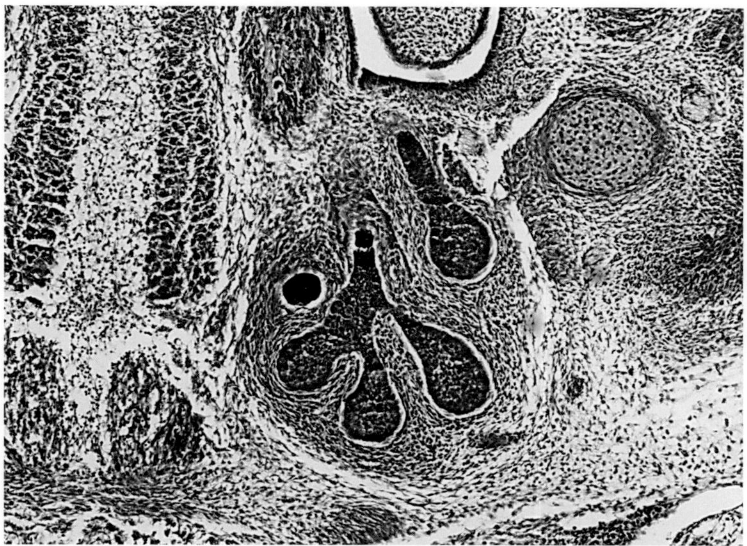

Fig. 1. Paraffin wax embedded section of an E13.5 mouse submandibular gland. The epithelial branches are solid at this early stage of development.

formation as salivary gland development proceeds can be examined in histological sections, and has been studied thoroughly in the murine submandibular gland [1]. These studies have revealed that small, separated lumina, surrounded by a multi-layer of cuboidal epithelial cells, begin to form along the centre of the previously solid presumptive ducts of the developing gland during the mid stages of branching morphogenesis at approximately embryonic day (E) 15.5 (fig. 1, 2). This stage of salivary gland development has consequently been termed the canalicular stage. By the late canalicular stage, these lumina are larger and are surrounded in places by a bilayer of cuboidal cells. Lumina also begin to appear in the end buds (the presumptive acini) at this stage. Subsequently, by the terminal differentiation stage of development at approximately E17.5, the lumina have expanded further and have become continuous in places, appearing wider and longer and now lined by a single epithelial layer in some areas with mesenchymal cells adjacent to their basal surfaces (fig. 3). The mature lumen eventually becomes fully functional shortly after birth.

What molecular and cellular events take place to create this tubular structure? Studies of other branched organs have revealed that there are a variety of biological ways to achieve a similar tubular form. In the mammary gland, initially solid epithelial branches are shaped into tubes by polarization of the outer epithelium followed by programmed cell death (apoptosis) and phagocytosis of central cells [2]. In contrast, in the rat lung, lumen formation has been shown to be driven

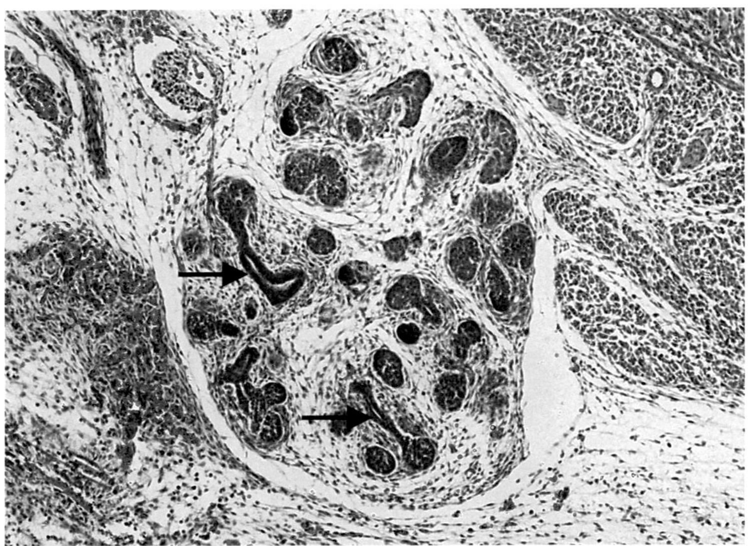

Fig. 2. Paraffin wax embedded section of an E15.5 mouse submandibular gland. Lumina are beginning to form in the presumptive ducts (arrows).

by an extension of the tracheal lumen into the lung bud with a sheet of polarized epithelial cells lining the lumen throughout lung development [3]. In the salivary gland, the process of lumen formation is thought to more closely resemble that of the mammary gland, involving two complementary processes: (i) the creation of a luminal space by programmed cell death and (ii) polarization of the epithelium lining the lumen.

Creating a Space

Apoptosis plays an important role in the development of many tissues. In the limb, apoptosis is involved in removing the interdigital cells to allow for development of distinct digits [4]. In the snake fang, apoptosis is instrumental in hollowing out the tooth, creating a tubular structure essential for venom delivery [5, also see Tucker, pp. 21–31]. Furthermore, many organs utilize apoptosis as a developmental mechanism including kidney [6], heart [7] and brain [8]. Evidence that apoptosis is involved in the creation of lumina in the salivary gland is derived from experiments using antibody labelling techniques to detect the DNA strand breaks or denaturation characteristic of apoptosis in mouse submandibular gland sections [9]. These techniques have shown that apoptotic cells are localized in the developing ducts at the canalicular stage of development at sites of current or future lumen formation. At the terminal differentiation stage, a reduction in

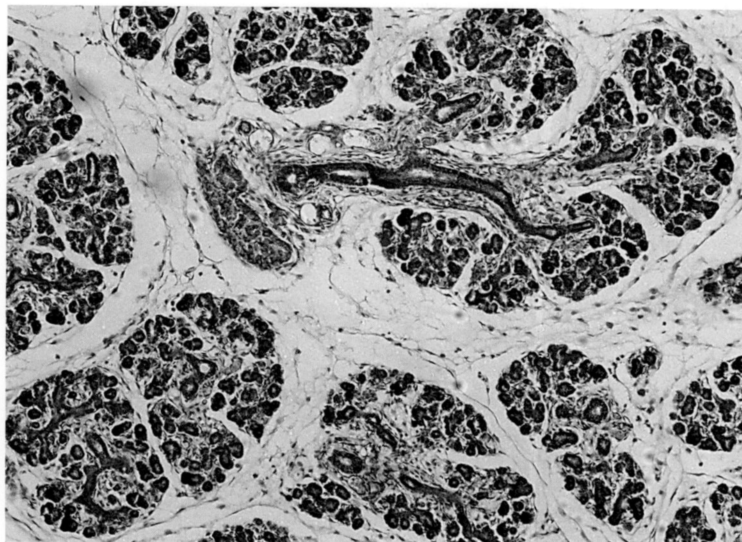

Fig. 3. Paraffin wax embedded section of an E17.5 mouse submandibular gland showing mature ductal and acinar lumina.

apoptosis is observed in areas where mature ductal lumina have formed, whereas apoptosis continues to be observed in the end buds where lumina develop slightly later [1].

Apoptosis is mediated by distinct pathways that can be initiated from outside the cell (the death receptor or extrinsic pathway) or inside the cell (the mitochondrial or intrinsic pathway) (fig. 4). Both pathways involve the activation of a family of cysteine proteases known as caspases. The caspase machinery acts as a cascade, beginning with the initiator caspases, such as caspases 8 and 9, which cleave and activate the downstream effector caspases 3, 6 and 7 to ultimately digest cellular components. In the intrinsic pathway, mitochondrial permeability increases in response to an intracellular apoptotic signal instigating the release of apoptotic proteins, including cytochrome C. Cytochrome C associates with apoptosis protease activating factor 1 (APAF1) and pro-caspase 9 to form a structure known as the apoptosome [10]. This leads to activation of caspase 9, which in turn activates downstream effector caspases. Conversely, the extrinsic pathway involves initiation of apoptosis via cell surface death receptors. These receptors belong to the tumour necrosis factor (TNF) superfamily and can bind a variety of apoptotic ligands. On ligand binding, the receptor's intracellular death domain forms a complex with the intracellular adaptor proteins Fas-associated protein with death domain (FADD) and TNFRSF1A-associated via death domain (TRADD), as well as pro-caspase 8. This induces the activation of caspase 8, which stimulates downstream activation of effector caspases.

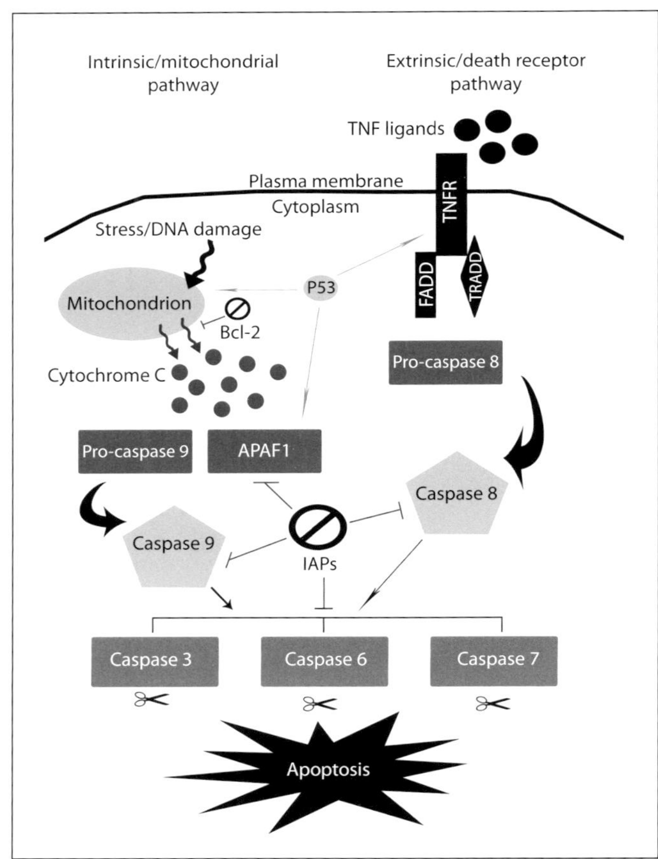

Fig. 4. Schematic of the intrinsic and extrinsic apoptosis pathways.

A number of the molecules involved in apoptotic signalling have been localized in the developing salivary glands, lending further weight to the hypothesis that apoptosis is involved in the formation of lumina. Immunohistochemical labelling of mouse submandibular gland sections has shown that activated caspase 3, a key effector caspase in both the intrinsic and extrinsic apoptosis pathways, is weakly detected in the developing branches at the pseudoglandular stage of development, but more strongly detected in the canalizing ducts at the canalicular stage [11]. This provides strong evidence that the apoptosis taking place in the developing ducts is caspase-dependent. Interestingly however, caspase 3 is not detected in the canalizing end buds, suggesting that caspases mediate duct canalization, but perhaps not presumptive acini lumen formation. Further evidence suggests that end bud as well as ductal lumen formation may involve p53, a tumour-suppressing transcription factor that can initiate apoptosis at a variety of points in the pathway, since p53 is present in the canalizing ducts and end buds of the mouse submandibular gland [12]. Extrinsic apoptosis signalling may be particularly important in lumen formation; pro-caspase 8 and the death receptor TNF-R1 are immunodetected in association with forming lumina [1]. Also, mRNA quantitative analysis of canalicular stage mouse submandibular glands has revealed a stage specific increase in expression of the death receptor ligand TNF, the intracellular adaptor protein TRADD and IL-6, a cytokine which is upregulated in response to TNF receptor binding [11]. However, TNF signalling via TRADD is also associated with cell survival and proliferative responses through the NFκB pathway, so the story remains far from lucid.

If it is the case that the central ductal and end bud cells in the developing salivary gland undergo apoptosis to leave a luminal space, what then prevents the remaining epithelial cells from also succumbing to this fate? Apoptosis in vivo is strictly controlled by a number of endogenous caspase inhibitors: proteins of the inhibitors of apoptosis (IAP) family. One such family member, survivin, is detected in the cytoplasm of epithelial cells during the pseudoglandular stage of mouse submandibular gland development, but at the canalicular stage is found in the nuclei of epithelial cells bordering the forming lumen while remaining in the cytoplasm of non-lumen-bounding epithelia [13]. Survivin has been shown to inhibit caspases only when translocated to the nucleus, and therefore may act as an inhibitor of apoptosis from the canalicular stage of salivary gland development, restricting loss of cells to those at the centre of the ducts. The anti-apoptotic proteins Bcl2, NfκB and receptor interacting protein (RIP), an intracellular adaptor protein associated with TNF signalling upstream of NfκB, are also detected in the epithelial cells bordering the developing lumen [11], and therefore may also be involved in protecting these cells from an apoptotic fate.

Cellular attachments are also likely to play an important role in apoptosis protection. The outermost layer of epithelial cells facing the basement membrane is thought to benefit from the unique protection from apoptosis conferred by a strong attachment to the extracellular matrix (ECM) [12]. The remaining epithelial cells not associated with the ECM are bound together only by weaker associations termed adherens junctions. It has been hypothesized that a cascade beginning with an increase in compression force on centrally located cells, brought about by an increase in proliferation of the surrounding cells and the 'doomed' cells' unique central position, initiates TNF signalling and an upregulation of IL-6. This cytokine has been shown to destabilize adherens junctions, which would leave the central cells without anchorage and prone to apoptosis [12]. In vitro mouse submandibular gland explant studies using siRNA or a function-blocking antibody to inhibit adherens junction formation (via the adherens junction protein E-cadherin) resulted in an increase in apoptosis in the developing ducts and dilated lumina [14].

Lining the Lumen: Cell-Cell Junctions

Clearance of a space alone is not sufficient to form a lumen; the formation of junctions between the outer luminal cells must accompany the space-clearing

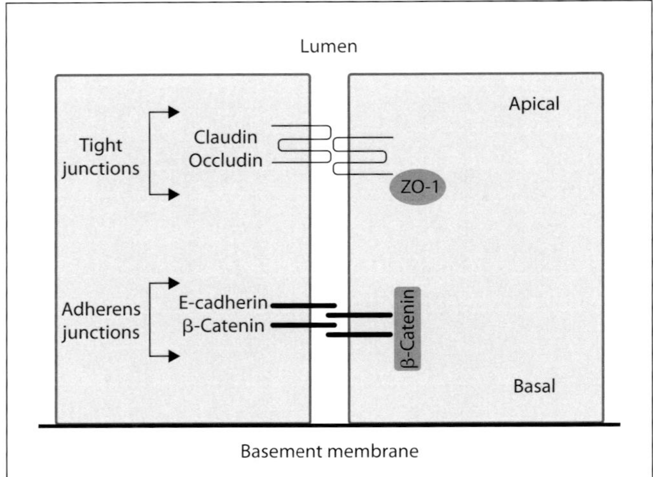

Fig. 5. Schematic of a submandibular gland lumen, showing localisation of junction proteins.

mechanism. The previous section introduced the concept that such cell-cell junctions are involved in apoptosis protection during lumen formation. In fact, these junctions are also thought to serve two additional vital purposes in salivary gland lumen development: (i) creation of a paracellular barrier around the lumen, controlling the diffusion of water and ions through the paracellular space, and (ii) blocking the movement of integral membrane proteins between apical and basolateral cell surfaces, allowing apical surfaces to become specialized for secretion and saliva modification.

The two main types of junctional complex present in salivary gland luminal epithelium are (1) tight junctions and (2) adherens junctions. Here we will describe these types of cell-cell junction in turn, describing their distribution in the developing salivary glands, and discussing their role in maintaining the paracellular barrier in salivary gland lumen-lining epithelium. Finally, we will consider the involvement of cell-cell junctions in establishing apicobasal polarity in the developing lumen.

Tight Junctions

Tight junctions are bonds between the basolateral membranes of adjacent cells, formed through the association of integral membrane proteins (fig. 5). The main proteins involved in tight junction formation are claudin and occludin. Both claudin and occludin exclusively localize at tight junctions, spanning the plasma membrane four times to create two extracellular loops which are involved in cell adhesion [15]. While claudins are a large family of molecules consisting of more than 20 members, and are small proteins (with a molecular mass of around 23 kDa), occludin has only one isoform and is larger, with a molecular mass of around 65 kDa. Occludin has specifically been shown to associate with the cytoplasmic scaffolding protein ZO-1 which is localized in the immediate vicinity of areas of the plasma membrane forming tight junctions [16].

On its journey from acini to oral cavity, the primary saliva secreted by the acini passes through the salivary ducts where it undergoes modification through the absorption and secretion of ions to create hypotonic saliva. The process of saliva modification in the ducts is strictly controlled, therefore an important characteristic of ductal luminal epithelium is its impermeability to the diffusion of solutes through the paracellular space. The acinar epithelium, on the other hand, must be permeable to water to provide fluid for saliva production. Tight junctions are thought to be vital contributors to ductal impermeability and the controlled permeability of the acini. Therefore, tight junctions are characteristic of adult salivary gland lumen lining epithelium, and the localization of claudins and occludin has been demonstrated in the mature lumen lining cells of rat salivary glands [17], mouse submandibular gland [18] and human salivary glands [19].

During early salivary gland development, prior to the initiation of branching morphogenesis, tight junction proteins such as claudins, occludin and ZO-1 are not present [20]. It is only later in development, at approximately E14, when end buds and branches begin to form, that tight junction proteins are first localized in the developing gland. By the late canalicular stage, at approximately E16, tight junction proteins are present in all the lumen-bounding epithelial cells [18, 20]. The temporal and spatial localization of tight junction proteins during salivary gland development suggests that tight junctions are required for paracellular barrier integrity, but functional evidence to support this supposition is only just beginning to accumulate. It has been shown that a submandibular gland epithelial cell line expressing GFP-fused claudin-4 has a higher transepithelial electrical resistance and a lower permeability than wild-type cells, indicating that at least this claudin isoform may play a role in tight junction barrier formation [21].

An interesting question is how differential expression of tight junction protein isoforms in the salivary gland contributes to the development and maintenance of the lumen. Immunohistochemistry in rat salivary gland sections has shown that although occludin is ubiquitous throughout the lumen-lining ductal epithelium, the localization of different claudin isoforms varies throughout the gland. Claudin-3 is present in the acinar cells and intercalated ducts, while claudin-4 is principally localized to the striated and interlobular ducts [17]. In human salivary glands, while claudin-1 is found in ductal, endothelial and some serous cells, claudins-2, -3, and -4 are found in both ductal and acinar cells, whereas claudin-5 is localized to endothelial cells only [19]. It seems likely that tight junction specificity reflects the different requirements for the control of paracellular movement of water and solutes in different areas of the gland; ductal epithelium must be impermeable to water and solutes, whereas the acinar epithelium must be permeable [22, 23]. For example, claudin-1 is thought to be particularly important in forming a tight paracellular barrier, since the Claudin-1

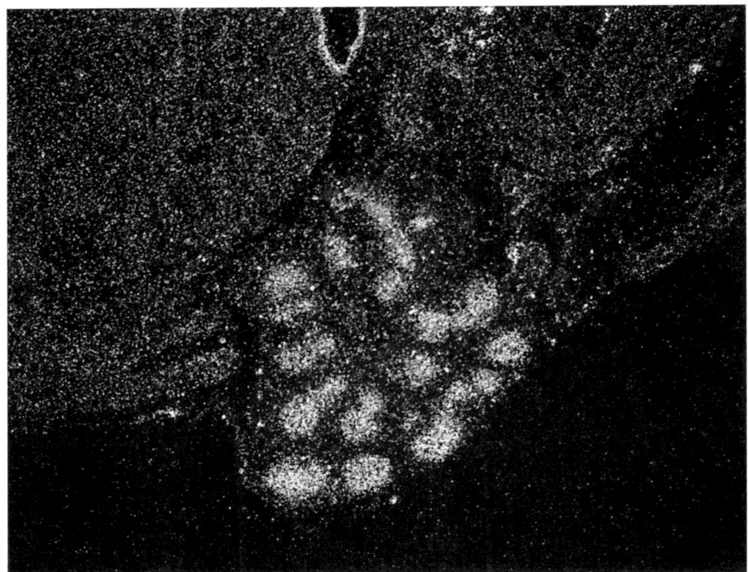

Fig. 6. Radioactive in situ hybridization on section through E14.5 mouse submandibular gland. White grains indicate E-cadherin expression confined to the epithelial portion of the developing gland.

knockout mouse dies at birth due to increased epidermal permeability [24]. Claudin-2, on the other hand, is thought to contribute to the weakening of tight junctions [25]. Also, tight junction proteins are able to bind signalling molecules [26] and it has been proposed that tight junctions are involved in the regulation of proliferation and differentiation [19]. Therefore another tempting hypothesis is that tight junction specificity in different areas of the salivary gland relates to specific signalling requirements, but this remains to be investigated.

Adherens Junctions

Adherens junctions also bind adjacent cells together via integral basolateral membrane proteins, but are generally thought to be weaker associations than tight junctions and are positioned more basally in the cell membrane (fig. 5). The key protein involved in adherens junction formation is epithelial (E)-cadherin, a member of the cadherin family of integral membrane proteins that mediate cell-cell adhesion in a calcium-dependent manner [27]. Adherens junctions are defined by their association with the actin cytoskeleton, and are formed when E-cadherins on adjacent cells form complexes with each other. Via its intracellular domain, E-cadherin binds its key regulators p120-catenin and β-catenin. It is thought that the role of β-catenin is to mediate the physical association of E-cadherin to the actin cytoskeleton, whereas p120 plays a role in the strength of cell-cell adhesion [28].

Consistent with their requirement for maintaining cell-cell adhesion in epithelial tissues, adherens junction proteins are present in salivary gland epithelium from early stages of development through to adulthood (fig. 6). Immunostaining of mouse submandibular gland sections has shown that both E-cadherin and β-catenin are distributed along the cell periphery of the submandibular gland epithelium from as early as E12, becoming localized to the lumen-lining epithelium later on [20]. The paramount importance of adherens junctions in salivary gland lumen development has been demonstrated in vivo in a conditional knockout mouse in which p120 is deleted in the submandibular gland epithelium, destabilizing adherens junction formation. The salivary glands of these mice exhibit lumina occluded by epithelial masses reminiscent of a pre-cancerous intraepithelial neoplasia, showing that stable adherens junctions are required for the proper development and maintenance of salivary gland lumina. Surprisingly, the submandibular glands of these mice exhibit another striking phenotype; the gland is comprised solely of ducts, and acinar differentiation does not take place [29]. Given that adherens junctions are localized in developing ducts as well as acini, this phenotype seems unexpected, and indicates that our knowledge of the role of adherens junctions in salivary gland development is far from complete.

Apicobasal Polarity

As primary saliva moves through the lumina of the salivary gland and undergoes ionic modification, ionic exchange takes place at the apical membrane. For this process to take place, lumen lining cells must therefore maintain strict apicobasal polarity. It is thought that tight junctions and adherens junctions play a major role in setting up this polarity by blocking the movement of integral membrane proteins between apical and basolateral cell surfaces, allowing each surface to become specialized for secretion.

Functional studies are beginning to demonstrate the importance of cell-cell junctions in the development of apicobasal polarity in the salivary gland. The expression of the polarity markers Crb3 (apical), NaK-ATPase (basolateral) and p63 (basal) requires proper adherens junction formation in developing mouse submandibular gland, since distribution of these markers is disrupted in the p120 knockout mouse in which adherens junctions fail to form correctly [29]. Tight junctions are similarly important for the ability of the cell to set up apicobasal polarity and therefore to secrete; occludin-deficient mice have been shown to lack characteristic secretory granules in the striated duct epithelial cells of the salivary glands [30].

Other studies have concentrated on identifying the signalling pathways upstream of the establishment of polarity in SG epithelium. There is evidence that this aspect of salivary gland development is controlled by the mesenchyme, shown by the fact that salivary gland epithelium combined with lung mesenchyme does not undergo polarization [31]. Signals within the epithelium of the developing gland also appear to be involved in polarization; recent work has shown that murine submandibular gland explants treated with exogenous sonic hedgehog protein, which is expressed in the epithelium during mouse submandibular gland development in vivo [32], show increased lumen formation and presence of cell-cell adhesion proteins [33].

Future Research

The study of lumen formation in developing organs is a worthwhile area of research. Not only is tube formation an intriguing aspect of organogenesis, it is also important in terms of understanding diseases of branching organs; for example, lumen filling is a hallmark of mammary gland carcinoma. Therefore, research into the mechanisms of lumen formation in the salivary glands will continue to be of value, and there are many unanswered questions. For example, although the localization of cells undergoing programmed cell death in the developing salivary glands suggests apoptosis as a mechanism for space clearance in lumen development, functional evidence to confirm this hypothesis is currently lacking. Furthermore, cell-cell junction proteins have been localised in the salivary gland luminal epithelium, however functional evidence to demonstrate their specific roles in lumen development, and to identify the signalling pathways upstream of the establishment of epithelial polarity in the developing gland, is only just beginning to accumulate. It is hoped that research over the next few years will fill these gaps in our knowledge.

References

1 Jaskoll T, Melnick M: Submandibular gland morphogenesis: stage-specific expression of TGF-α/EGF, IGF, TGF-β, TNF, and IL-6 signal transduction in normal embryonic mice and the phenotypic effects of TGF-β_2, TGF-β_3, and EGF-R null mutations. Anat Rec 1999;256:252–268.

2 Reginato MJ, Muthuswamy SK: Illuminating the center: mechanisms regulating lumen formation and maintenance in mammary morphogenesis. J Mammary Gland Biol Neoplasia 2006;11:205–211.

3 Burri PH, Moschopulos M: Structural-analysis of fetal-rat lung development. Anat Rec 1992;234:399–418.

4 Zakeri Z, Quaglino D, Ahuja HS: Apoptotic cell-death in the mouse limb and its suppression in the hammertoe mutant. Dev Biol 1994;165:294–297.

5 Zahradnicek O, Horacek I, Tucker AS: Viperous fangs: development and evolution of the venom canal. Mech Dev 2008;125:786–796.

6 Chandler D, Elnaggar AK, Brisbay S, Redline RW, McDonnell TJ: Apoptosis and expression of the bcl-2 protooncogene in the fetal and adult human kidney – evidence for the contribution of bcl-2 expression to renal carcinogenesis. Hum Pathol 1994;25:789–796.

7 James TN: Normal and abnormal consequences of apoptosis in the human heart – from postnatal morphogenesis to paroxysmal arrhythmias. Circulation 1994;90:556–573.

8 Yoshida H, Kong YY, Yoshida R, Elia AJ, Hakem A, Hakem R, Penninger JM, Mak TW: APAF1 is required for mitochondrial pathways of apoptosis and brain development. Cell 1998;94:739–750.

9 Tucker AS: Salivary gland development. Semin Cell Dev Biol 2007;18:237–244.

10 Zou H, Li YC, Liu HS, Wang XD: An APAF-1 center dot cytochrome C multimeric complex is a functional apoptosome that activates procaspase-9. J Biol Chem 1999;274:11549–11556.

11 Melnick M, Chen HM, Zhou YM, Jaskoll T: Embryonic mouse submandibular salivary gland morphogenesis and the TNF/TNF-R1 signal transduction pathway. Anat Rec 2001;262:318–330.
12 Melnick M, Jaskoll T: Mouse submandibular gland morphogenesis: a paradigm for embryonic signal processing. Crit Rev Oral Biol Med 2000;11:199–215.
13 Jaskoll T, Chen H, Min Zhou Y, Wu D, Melnick M: Developmental expression of survivin during embryonic submandibular salivary gland development. BMC Dev Biol 2001;1:5.
14 Walker JL, Menko AS, Khalil S, Rebustini I, Hofftnan MP, Kreidberg JA, Kukuruzinska MA: Diverse roles of E-cadherin in the morphogenesis of the submandibular gland: insights into the formation of acinar and ductal structures. Dev Dyn 2008; 237:3128–3141.
15 Furuse M, Hirase T, Itoh M, Nagafuchi A, Yonemura S, Tsukita S: Occludin – a novel integral membrane-protein localizing at tight junctions. J Cell Biol 1993;123:1777–1788.
16 Furuse M, Itoh M, Hirase T, Nagafuchi A, Yonemura S, Tsukita S: Direct association of occludin with ZO-1 and its possible involvement in the localization of occludin at tight junctions. J Cell Biol 1994; 127:1617–1626.
17 Peppi M, Ghabriel MN: Tissue-specific expression of the tight junction proteins claudins and occludin in the rat salivary glands. J Anat 2004;205:257–266.
18 Hashizume A, Ueno T, Furuse M, Tsukita S, Nakanishi Y, Hieda Y: Expression patterns of claudin family of tight junction membrane proteins in developing mouse submandibular gland. Dev Dyn 2004;231:425–431.
19 Maria OM, Kim JWM, Gerstenhaber JA, Baum BJ, Tran SD: Distribution of tight junction proteins in adult human salivary glands. J Histochem Cytochem 2008;56:1093–1098.
20 Hieda Y, Iwai K, Morita T, Nakanishi Y: Mouse embryonic submandibular gland epithelium loses its tissue integrity during early branching morphogenesis. Dev Dyn 1996;207:395–403.
21 Michikawa H, Fujita-Yoshigaki J, Sugiya H: Enhancement of barrier function by overexpression of claudin-4 in tight junctions of submandibular gland cells. Cell Tissue Res 2008;334:255–264.
22 Anderson JM: Molecular structure of tight junctions and their role in epithelial transport. News Physiol Sci 2001;16:126–130.
23 Tsukita S: Tight junctions and epithelial cell polarization/growth. J Cancer Res Clin Oncol 2001; 127:S3.
24 Furuse M, Hata M, Tsukita S: Mice lacking claudin-1, a constituent of tight junction strands. Mol Biol Cell 2001;12:725.
25 Stevenson BR, Anderson JM, Goodenough DA, Mooseker MS: Tight junction structure and ZO-1 content are identical in two strains of Madin-Darby canine kidney cells which differ in transepithelial resistance. J Cell Biol 1988;107:2401–2408.
26 Mitic LL, Anderson JM: Molecular architecture of tight junctions. Annu Rev Physiol 1998;60:121–142.
27 Suzuki S, Sano K, Tanihara H: Diversity of the cadherin family – evidence for eight new cadherins in nervous tissue. Cell Regul 1991;2:261–270.
28 Davis MA, Ireton RC, Reynolds AB: A core function for p120-catenin in cadherin turnover. J Cell Biol 2003;163:525–534.
29 Davis MA, Reynolds AB: Blocked acinar development, E-cadherin reduction, and intraepithelial neoplasia upon ablation of p120-catenin in the mouse salivary gland. Dev Cell 2006;10:21–31.
30 Saitou M, Furuse M, Sasaki H, Schulzke JD, Fromm M, Takano H, Noda T, Tsukita S: Complex phenotype of mice lacking occludin, a component of tight junction strands. Mol Biol Cell 2000;11:4131–4142.
31 Iwai K, Hieda Y, Nakanishi Y: Effects of mesenchyme on epithelial tissue architecture revealed by tissue recombination experiments between the submandibular gland and lung of embryonic mice. Dev Growth Differ 1998;40:327–334.
32 Jaskoll T, Leo T, Witcher D, Ormestad M, Astorga J, Bringas P, Carlsson P, Melnick M: Sonic hedgehog signaling plays an essential role during embryonic salivary gland epithelial branching morphogenesis. Dev Dyn 2004;229:722–732.
33 Hashizume A, Hieda Y: Hedgehog peptide promotes cell polarization and lumen formation in developing mouse submandibular gland. Biochem Biophys Res Commun 2006;339:996–1000.

Kirsty L. Wells
Department of Craniofacial Development, 27th Floor, Guy's Tower
Guy's Campus, King's College London, London SE1 9RT (UK)
Tel. +44 207 188 7384, Fax +44 207 188 1674
E-Mail kirsty.wells@kcl.ac.uk

Tucker AS, Miletich I (eds): Salivary Glands. Development, Adaptations and Disease.
Front Oral Biol. Basel, Karger, 2010, vol 14, pp 90–106

Epithelial Stem/Progenitor Cells in the Embryonic Mouse Submandibular Gland

Isabelle. M.A. Lombaert · Matthew. P. Hoffman

National Institute of Dental and Craniofacial Research, National Institutes of Health Bethesda, Md., USA

Abstract

Salivary gland organogenesis involves the specification, maintenance, lineage commitment, and differentiation of epithelial stem/progenitor cells. Identifying how stem/progenitor cells are directed along a series of cell fate decisions to form a functional salivary gland will be necessary for future stem cell regenerative therapy. The identification of stem/progenitor cells within the salivary gland has focused on their role in postnatal glands and little is known about them in embryonic glands. Here, we have reviewed the information available for other developing organ systems and used it to determine whether similar cell populations exist in the mouse submandibular gland. Additionally, using growth factors that influence salivary gland epithelial morphogenesis during development, we have taken a simple experimental approach asking whether any of these growth factors influence early developmental lineages within the salivary epithelium on a transcriptional level. These preliminary findings show that salivary epithelial stem/progenitor populations exist within the gland, and that growth factors that are reported to control epithelial morphogenesis may also impact cell fate decisions. Further investigation of the signaling networks that influence stem/progenitor cell behavior will allow us to hypothesize how we might induce autologous stem cells to regenerate damaged salivary tissue in a therapeutic context.

Submandibular gland (SMG) organogenesis involves inductive interactions between mesenchymal and epithelial cells. Recent research on SMG development has focused on the cross-talk between mesenchymal and epithelial cells that control branching morphogenesis [1–6]. The SMG mesenchyme plays an instructive role and produces many growth factors that regulate epithelial morphogenesis by controlling processes such as proliferation, differentiation, migration, and cell death. During epithelial morphogenesis, primitive stem/progenitor cells also undergo a series of cell fate decisions that give rise to more differentiated cell types while simultaneously maintaining a reservoir of stem/progenitor cells. Therefore, a major challenge in the field is to identify epithelial salivary gland stem/progenitor cells and determine which signaling pathways direct their cell fate decisions along different cell lineages. Here, we will provide background on SMG development that will offer insight into stem/progenitor cells, and then review what is known about adult salivary gland stem/progenitor cells, focusing on some of the known stem/progenitor cell markers identified in other developing organ systems. Since there are no reports identifying stem/progenitor cells in embryonic SMGs, we have included some preliminary data: microarray and qPCR analyses that begin to define what types of salivary stem/progenitor cells exist in the SMG, and experiments investigating whether growth factors influence epithelial stem/progenitor cell fate.

By definition, a stem cell is capable of both unlimited self-renewal and differentiation into all mature gland cell types. As stem cells differentiate into more committed progenitor cells, they lose their self-renewing capacity and become more restricted to one cell type lineage. These progenitor cells, also termed transit amplifying (TSA) cells, are highly proliferative and can give rise to multiple differentiated cell types. Alternatively, a stem cell might produce one daughter cell that immediately commits to a differentiated cell without the need for further cell division or for a TSA cell [7]. However, there is currently no clear distinction between stem cells and TSA progenitor cells in the SMG. Therefore, we have defined all primitive cells in this review as stem/progenitor cells. Nevertheless, in both cell types, differentiation occurs by either silencing or activating gene transcription. Genes involved in self-renewal will be highly expressed in the stem/progenitor cells and become undetectable in differentiated cells. Genes with low expression in the stem/progenitor cells may increase their expression in one or both matured daughter cells as they differentiate. These cell fate decisions are regulated in a spatiotemporal manner during gland development and are controlled by intrinsic signals and/or by the environment, which includes growth factors, cell-cell interactions, and the extracellular matrix.

As reviewed in other chapters of this book, the SMG initiates at approximately 11 days post-coitum (E11), when an ectoderm-derived oral epithelium interacts with the neural crest-derived mesenchyme, forming an epithelial placode. The epithelium invaginates into the mesenchyme by E12, where an end bud enlarges on a stalk of epithelium. At E13, clefts form on the enlarged end bud and branching morphogenesis begins, along with lumen formation (fig. 1). At this stage, two distinct epithelial cell types, the primary duct and the end buds, are morphologically observed during dissection under a light microscope and can be easily separated and analyzed. Major cell differentiation occurs at E15, followed by postnatal gland maturation [8–12]. Classic experiments using heterotypic tissue recombination have provided insight into the tissue patterning and the plasticity of the epithelial progenitor cells as they respond to different inductive mesenchymal signals [1, 3]. The branching pattern is dependent on the source of the organ-specific mesenchyme so that lung, mammary, and pituitary epithelia develop a 'salivary-like' branching morphology when combined with embryonic salivary mesenchyme [2, 4–6]. However, SMG epithelial growth could only be induced by the addition of either urogenital or SMG mesenchyme. These data suggest that specific SMG mesenchyme-secreted growth factors are needed when

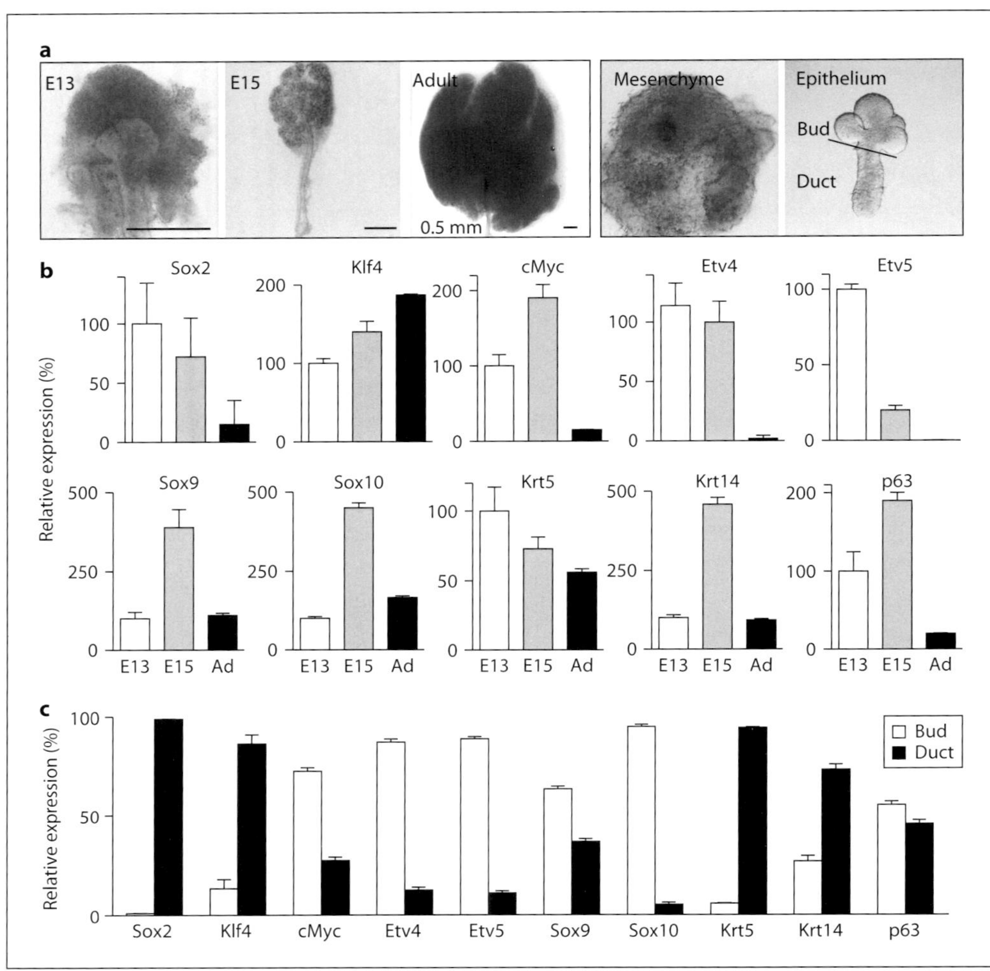

Fig. 1. Gene expression analysis of stem/progenitor cell-related markers in the developing SMG. **a** Branching morphogenesis of the mouse submandibular gland (SMG) begins at E13 as the end bud clefts. At E15, cell differentiation occurs, which is followed by postnatal cell maturation to form the adult gland. E13 glands were mechanically separated into a mesenchyme and epithelial compartment. The latter was further mechanically separated into an end bud and duct compartment. **b** Relative gene expression analysis of markers related to embryonic stem cell maintenance *(Oct3/4, Nanog, Sox2, Klf4, cMyc)*, stem/progenitor cell differentiation/self-renewal *(Etv4, Etv5, Sox9, Sox10)*, and basal stem/progenitor cells *(Krt5, Krt14, p63)* in E13, E15, and Adult (Ad) SMGs. The expression of *Oct3/4* and *Nanog* were not detected by qPCR. *Sox2, Etv4, Etv5* and *Krt5* decrease during development, while *Klf4* increases towards adulthood. Other genes such as *cMyc, Sox9, Sox10, Krt14* and *ΔNp63* are present throughout development, with maximum expression at E15. **c** The relative gene expression was compared between the E13 epithelial end buds and ducts. Ducts express higher levels of *Sox2* and *Klf4* and *Krt5*. In contrast, *cMyc, Etv4, Etv5, Sox9,* and *Sox10* were detected in the end bud compartment. *Krt14* and *ΔNp63* were expressed in both the duct and end bud compartments. Agilent whole mouse genome microarray analysis was performed on triplicate pooled biological samples; gene expression was normalized to the level of expression at E13. A probe for *Oct3/4* is not present on the Agilent microarray and the expression of *Nanog* was low. Therefore, qPCR was used to detect their gene expression, which was normalized to the housekeeping gene *29S*. Error bars represent SEM.

evaluating SMG epithelial cell fate decisions. Importantly, the capacity of the mesenchyme to re-imprint the epithelial cell fate only occurs with early epithelia (< E16), suggesting that later stages of development have less stem/progenitor cells and more committed differentiated cells. Consequently, we have focused our studies on identifying stem/progenitor cell populations present early in development by evaluating E13 SMG epithelium.

Important information regarding genes that are critical for SMG epithelial stem/progenitor cell survival and maintenance comes from the analysis of genetically modified mice [reviewed in 13]. Studies of mice lacking fibroblast growth factors have been particularly instructive. Conditional loss of fibroblast growth factor 8 (FGF8) in the ectoderm-derived epithelium results in severe gland hypoplasia; only a rudimentary epithelial bud develops [14], and loss of its mesenchymal receptor FGFR2c produces a similar effect [15, 16]. FGF8 signaling modulates both mesenchymal FGF10 and epithelial sonic hedgehog signaling, which explains the severe phenotype with the loss of epithelial FGF8. Salivary gland agenesis further occurs with the loss of either FGFR2b or its mesenchymal-derived ligand FGF10, although a single hypoplastic end bud does initially form [15, 17]. Similar effects occur in p63 null mice, with p63 normally being expressed in the basal stem/progenitor layers of many ectodermal organs [18]. As such, FGF8 and FGF10 signaling, and p63, are critical for the survival and growth of epithelial stem/progenitor cells. In addition, epidermal growth factor receptor (EGFR)-null mice have SMG hypoplasia, but differentiation appears normal [19]. This suggests that the number of progenitor cells may be reduced, resulting in a smaller gland. However, direct analyses of the stem/progenitor populations in these genetically modified mice have not been performed.

Adult Salivary Gland Stem/Progenitor Cells

Interest in identifying progenitor cells in the adult gland has arisen from the potential therapeutic application of regenerating salivary tissue after therapeutic irradiation of head and neck tumors, or for replacing damaged glands with bioengineered artificial tissue [20]. The adult gland is comprised of two major epithelial compartments – the ducts, which transport and modify saliva, and the acinar cells, which produce saliva. These cells are surrounded by a stromal matrix containing contractile myoepithelial cells, (myo)fibroblasts, immune cells, endothelial cells, and neurons. Adult stem cells could potentially be found in many different niches within an organ in a similar manner to skin stem cells, which reside in the interfollicular epidermis, hair follicles, sebaceous glands, and neural crest mesenchyme [reviewed in 21]. Therefore, multiple stem/progenitor cell types may reside within the SMG epithelium and mesenchyme, and a variety of techniques have been used to identify these.

Lineage tracing with genetically marked cells is the most direct genetic method of determining the progenitor cell populations in the gland. Ascl3 is a transcription factor localized in the duct cells of the salivary glands that increases in expression during postnatal development [22]. Recently, *Ascl3*-expressing SMG duct cells in mice were shown to form both duct and acinar cells by lineage tracing using a Cre-recombinase reporter system [23]. As such, a direct progenitor-progeny relationship was demonstrated between the multipotent Ascl3-expressing duct cells and the acinar cell compartment. Reversible duct ligation is another commonly used technique to identify stem/progenitor cells and is described in detail in the following chapter by Carpenter and Cotroneo [pp. 107–128]. Duct ligation causes atrophy of the gland with major acinar cell loss. Following the removal of the ligation, intercalated duct cells proliferate and reform acinar cells. This observation suggested intercalated duct cells were acinar progenitors [24, 25], which was further supported by acinar differentiation in vitro of duct cells from human, macaque [26] and rat [27] SMG/parotid glands.

Another way to identify stem/progenitor cells is to localize label-retaining cells (LRCs) in adult glands. These cells are slowly dividing and retain DNA-binding dyes, such as BrdU, long after a transient exposure. LRCs have been identified in acinar cells, ducts, myoepithelium, and connective tissue cells [24, 25, 28, 29], suggesting multiple progenitor cell types exist in different cell compartments in the adult salivary glands. Fluorescent-activated cell sorting (FACS) is a common and direct technique to label and isolate stem/progenitor cells from adult tissue. The initial characterization of cell surface markers that identify stem/progenitor cells from fresh or post-duct-ligated glands showed they were $Sca1^+$ and $cKit^+$ [30, 31], $\alpha_6\beta_1$-integrin$^+$ [32] or $CD49f^+$ (α6-integrin) and $Thy1^+$ [33, 34]. Cells isolated from neonatal rat and adult rat/human glands, termed salivary gland stem cells (SGSC), transdifferentiated into either pancreatic or hepatic cells [30, 31, 34, 35], and regenerated hepatectomized livers after transplantation. These SGSCs are similar to in vitro cultured pancreas stem cells [36], highlighting the plasticity of early organotypic stem cells, which suggests that similarities exist between different exocrine organs. SGSCs also express c-Kit, and the pluripotency markers Nanog and Oct3/4. However, this heterogeneous cell population has mesenchymal-like stem cell properties based on their cell surface $CD44^+$, $CD90^+$, and $CD105^+$ expression. Other stem/progenitor cells that have been characterized by FACS are the side population (SP) cells, which have a unique ABC-transporter channel that excludes toxic dyes such as rhodamine or Hoechst-33342. SP cells have been identified in adult salivary glands and make up 1% of the cell population, although their capacity to self-renew and differentiate has not been evaluated [37].

Many cell types mentioned in this section have been hypothesized to be salivary gland stem/progenitor cells but have not been tested for their capacity to regenerate damaged SMGs. However, epithelial c-Kit$^+$ duct cells isolated by FACS from adult mouse SMGs have been shown to functionally and morphologically regenerate irradiated adult mouse SMGs [38]. Autologous SMG stem cell transplantation is a clinical approach to potentially treat head and neck cancer patients suffering from radiation-induced hyposalivation. However, this heterogeneous c-Kit^+ cell population likely contains multiple cell types, including self-renewing stem/progenitor cells with multipotent differentiation capacities.

In summary, there are numerous examples in the literature of postnatal salivary gland cell types that exhibit self-renewing, transdifferentiating, and multipotent capacities. However, it remains to be determined from which stem/progenitor cell type they originate during embryonic gland development, and what influences their cell fate decisions as they develop along different cell lineages.

Embryonic Salivary Gland Stem/Progenitor Cells

To date, no lineage-tracing studies identifying early embryonic SMG stem/progenitor cells have been reported, and few molecular markers have been identified in the epithelial compartment of the embryonic gland. However, in other developing organ systems, specific transcription factors (TFs) have been used to identify stem/progenitor cell types from cells undergoing early cell fate decisions. These cell fate decisions are coordinated by TFs that regulate genes involved in self-renewal and differentiation. Additionally, stem/progenitor cells have also been characterized with cytoskeletal markers, particularly the intermediate filament keratins, to define their cellular differentiation state. Therefore, we analyzed whether TFs involved in stem cell self-renewal (Sox2, Klf4, Nanog, Oct3/4, cMyc, and the Etv family), progenitor differentiation (other Sox genes), and basal progenitor cell markers (cytokeratins) were expressed in the SMG.

We hypothesized that early epithelial cell fate decisions have occurred within the epithelium before branching morphogenesis begins (E13), and that TFs and cell markers involved in stem/progenitor cell self-renewal and early cell lineage specification will already be expressed in distinct epithelial cell compartments. As development and gland maturation proceeds and cells become committed to a particular cell lineage, the expression of genes involved in self-renewal and cell lineage specification would change. Therefore, the presence of these genes in the SMG was evaluated by genome-wide microarray analysis or qPCR at important time-points during development when major cell fate changes are likely occurring; at E13 as branching morphogenesis begins, at E15 when end bud differentiation starts, and in the fully functional adult gland (fig. 1b). For genes implicated in stem cell self-renewal, embryonic stem cells were used as a positive control. Furthermore, we analyzed their expression within the two morphologically distinct E13 epithelial compartments: the initial duct and end buds (fig. 1c). We also hypothesized that growth factors regulating SMG development would also regulate epithelial cell fate decisions. Therefore, we added growth factors that are important for SMG development to isolated E13 epithelium and measured the early (2 h) downstream transcriptional effects on the genes implicated in stem/progenitor cell self-renewal (fig. 2). Importantly, this is by no means a comprehensive

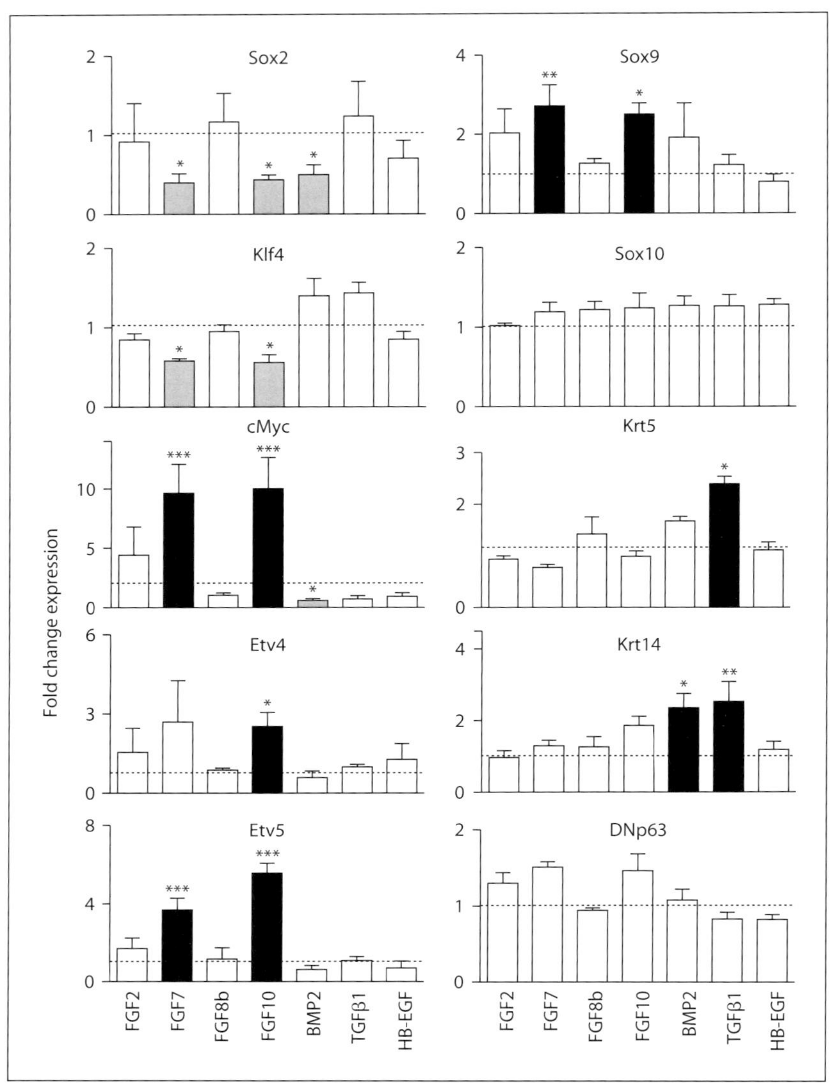

Fig. 2. Growth factors involved in SMG development regulate stem/progenitor cell-related gene expression in E13 SMG epithelium. Different growth factors produced by the SMG in vivo such as FGF2, FGF7, FGF8b, FGF10, BMP2/4, TGF-β1/2, and HBEGF were added to isolated SMG epithelium for 2 h. Both FGF10 and FGF7 downregulate *Sox2* and *Klf4*, which are mainly expressed in the duct. FGF7 and FGF10 also upregulate *cMyc*, *Etv5*, and *Sox9*, which are expressed in the end bud, and FGF10 specifically upregulates *Etv4*. BMP2 also downregulates *Sox2* and *cMyc* expression. Basal stem cell-related genes *Krt5* and *Krt14* are both upregulated by TGF-β1. BMP2 also increases *Krt14* expression. In contrast, *Sox10* and *ΔNp63* are not regulated by any of the growth factors. In addition, BMP4 (not shown) gave similar results to BMP2, and TGF-β2 (not shown) gave similar results to TGF-β1. Gene expression was normalized to the housekeeping gene *29S* and normalized to epithelia cultured in media alone for 2 h. At least five epithelial rudiments were used for each condition and the results of three independent experiments were combined. Mean ± SEM. Statistical analyses were performed using a one-way ANOVA or t test with * $p < 0.05$, ** $p < 0.01$, *** $p < 0.001$.

analysis of all the growth factors and genes involved, but provides an initial starting point for discussion.

SMG Expresses TFs Involved in Embryonic Stem Cell Self-Renewal

In 2006, a groundbreaking report showed that differentiated cells could be reprogrammed to become pluripotent by forced viral transduction of four TFs, Sox2, Klf4, cMyc, and Oct3/4 [39]. The cells were called induced pluripotent stem (iPS) cells, and exhibited morphological properties of embryonic stem (ES) cells, suggesting that these TFs were major regulators of stem cell maintenance. Unexpectedly, Nanog, another TF thought to be involved in ES cell maintenance, was dispensable for iPS formation. Nonetheless, these TFs have become the hallmark of stem cell self-renewal, but whether they are present in stem/progenitor cells within the SMG is unknown.

Oct3/4 was the first TF to be identified as a master regulator of pluripotency in ES cells. Apart from ES cells, Oct3/4 is present in adult stem cells such as bone marrow mesenchymal stem cells, prostate neuroendocrine cells [40], muscle-derived progenitors [41] and also in multiple cancers. Both Oct3/4 and Nanog expression were reported in cultured adult SGCS cells [36]. However, in our analyses, *Oct3/4* and *Nanog* expression were undetectable by qPCR at any stage of SMG development (fig. 1b). We used cDNA from ES cells to confirm that the primers amplified *Oct3/4* and *Nanog* with our qPCR conditions.

Sox2 is a TF expressed in many tissues during embryonic development including the pancreas, retina, brain [42–44], neural crest, tongue, pituitary gland [45], the lungs, tongue, and esophagus [46–49]. Generation of Sox2 null embryos demonstrated that it was required for epiblast and extraembryonic ectoderm formation. Conditional deletion of Sox2 in early tissue development results in abnormal differentiation of organs with characteristic loss of basal cells and differentiated lineages [50]. Consistent with Sox2 being a stem/progenitor cell marker, our data shows that *Sox2* expression gradually decreases from E13 towards adulthood (fig. 1b). At E13 it is specifically localized to the epithelial ducts (fig. 1c), suggesting that this region may be a source of stem/progenitor cells during gland development.

Klf4 acts as a tumor suppressor by inhibiting proliferation via p21 activation, but is also involved in cell differentiation of epithelium in the intestine, skin, lung, testis, and cornea. Klf4 increases during SMG development (fig. 1b), and our analysis at E13 reveals relatively more expression of Klf4 in the duct, similar to Sox2 (fig. 1c). Interestingly, Klf4 may regulate cytokeratin 19 (Krt19) expression [51], which we show to be expressed in the SMG ducts [52], and is known to be expressed by TSA cells in the pancreas [53].

cMyc is a basic helix-loop-helix TF involved in the maintenance of ES cells, in cell growth, differentiation, and proliferation. As a proto-oncogene, it promotes cell cycle progression of G1 into S phase and counteracts the anti-proliferative effect of Klf4 [54].

In the developing SMG, a relative increase in cMyc expression is detected at E15 (fig. 1b) when proliferation and gland expansion are rapidly occurring. Additionally, cMyc is expressed more in the E13 end buds than in the duct (fig. 1c), which is analogous to the cMyc$^+$ proliferating tip-precursor cells in the pancreas [55] and our previous reports that highly proliferative cells are present in the SMG end buds [56]. Thus, cMyc expression may indicate proliferating progenitor cells within the end buds.

Taken together, these data suggest that two distinct progenitor cell populations exist in the E13 SMG; the duct containing the more 'primitive' or pluripotent cell types compared to the end bud, i.e. higher relative expression of the TFs involved in ES cell maintenance, namely Sox2 and Klf4. Whereas, cMyc was more abundant in the end buds, consistent with the hypothesis that the end bud cells have gone through an early cell fate decision and are more proliferative. The observation that *Oct3/4* and *Nanog* were not detectable by E13 suggests that the SMG stem/progenitor cells present no longer have the same capacity to self-renew as ES cells.

ETS and Sox-Related Transcription Factors are Expressed in the Epithelial End Bud

ETS TFs are a large family of TFs associated with stem cell maintenance, cell proliferation, differentiation, and tumorigenesis [57, 58], and some of them may also function as transcriptional repressors. The maintenance of mesenchymal stem cells [59] requires Etv5 expression, and both Etv4 and Etv5 are expressed in proximal pancreatic progenitor cells that are regulated by mesenchyme-secreted FGF10 [60]. *Etv4* is maintained in the gland until E15 and decreases towards adulthood, whereas *Etv5* expression markedly decreases at E15 and is undetectable in the adult gland, suggesting it is only involved in early SMG development. In the E13 SMG, both TFs were highly expressed in the end bud compared to the duct. Similar to the pancreas, the SMG end bud is surrounded by an FGF10-producing mesenchyme [13]. Therefore, both Etv5 and Etv4 may be associated with the maintenance of putative end bud progenitor cells under the control of FGF10.

In addition to the ES cell-related Sox2, other Sox TFs also play a role in stem cell specification/maintenance in a variety of tissues, marking both early and late stem/progenitor cells. Sox9 is involved in the development of many organs, including the pancreas, pituitary gland, kidney, intestine, and gonad [61–63]. Sox9 is not critical for the initial formation of organotypic stem cells, but it is required for their specification, maintenance, survival, and proliferation [64]. However, it is the level of Sox9 expression that determines its role in stem/progenitor cell regulation. For example, in intestinal epithelium, Sox9 is abundant in post-mitotic Paneth progenitor cells but is expressed less in the highly proliferative crypt columnar stem cells. Increased levels of Sox9 repress cMyc and cyclin-D1-regulated proliferation and induce Paneth cell specification [65]. This is in contrast to the crypt stem cells where low Sox9 levels maintain the stem cells as they proliferate. Thus, Sox9 can have different effects depending on its level of expression in different types of stem/progenitor cells within a tissue, and this also occurs in the pancreas [66, 67], hair [68, 69], heart valve [70], cartilage [64], and neural crest [71, 72]. In the SMG, *Sox9* is expressed in the E13 epithelial end bud cells, suggesting it might exert a similar effect on the maintenance or early specification of the end bud cells as they proliferate.

Sox10 is also associated with stem/progenitor maintenance/specification. Human mutations in Sox10 cause the Waardenburg-Shah syndrome, characterized by hypopigmentation of the skin, heterochromia irides, deafness, and absence of enteric ganglia. Therefore, Sox10 is required for the development of multipotent neural crest cells and their derivatives, the Schwann cells, glia, and melanocytes. Sox10 can also have different effects depending on its level of expression in different stem/progenitor cells within a tissue. Before lineage segregation, low Sox10 expression increases neural crest progenitor cell survival, whereas later in development, higher Sox10 levels induce lineage determination of glial cells. During adulthood, Sox10 remains in glial cells only, suggesting it might be required to prevent differentiation into other neuronal cell lineages [73].

In E13 SMGs, the epithelial end buds express *Sox10* (fig. 1c). Sox10 expression was previously reported in E14.5 end bud cells and in several types of duct cells [74]. If Sox9 and Sox10 proteins have a similar function during SMG development as in neural crest and intestinal formation, our analyses suggest that SMG end bud progenitor cells have undergone some specification. This hypothesis is supported by the observation that *Sox9* and *Sox10* expression increases at E15, the time cell differentiation begins (fig. 1b). In addition, in adult human SMGs, Sox10 was located in myoepithelial cells [75]. Taken together, these data suggest that Sox10 might be involved in later basal/myoepithelial cell specification and maintenance. However, it is not known if they are co-expressed in the same cells in the early SMG or what their function is.

SMG Epithelia Contain Basal Progenitor Cells

An alternative approach to defining stem/progenitor lineages is by evaluating the cytoskeleton keratin profiles of cells. Keratins occur as heterodimer pairs of a type I and type II keratin, and are expressed in spatiotemporal and tissue-specific manners. Keratin mutations lead to disorders in several tissues such as the pancreas, liver, and intestine. The keratin expression patterns have been used to characterize basal, intermediate, and luminal cell types in many epithelial tissues. The basal Krt5$^+$ and Krt14$^+$ cell layer contains the niche for multiple tissue stem/progenitor cells in all multilayered epithelia such as prostate, mammary glands and skin bulge stem cells [76]. Krt5/Krt14 expression is downregulated during differentiation, and there is a transition to other keratin types as the cells move into suprabasal layers. Depending on the tissue, the cells transition to express keratins such as Krt1/10 in the epidermis and Krt8/Krt18 in luminal prostate and mammary gland cells. In the oral epithelium and mammary duct cells, basal cells first lose Krt14, which is replaced by Krt15, Krt17 and Krt19 [77]. Krt15 and Krt17 are limited to basal intermediate cells, while cells moving from the basal into the luminal layer only maintain Krt19. Further terminal differentiation leads to loss of Krt19 and increased Krt8/Krt18 expression [53]. Surprisingly, keratin profiles of the early SMG have not been thoroughly characterized. Krt8 staining (Troma-1 antibody) was observed throughout the E13 SMG epithelium, although it appeared higher in the ducts than the end buds [78]. Also, Krt7 has been used as a luminal duct cell marker in the E13.5 SMG [79]. We observed Krt5 expression in basal duct cells of the E13 SMG [52] as well as a subpopulation of cells within the end bud region. The expression pattern of Krt5 and Krt14 is different, with *Krt5* expression decreasing in the adult SMG (fig. 1b), whereas *Krt14* has increased expression at E15 and is similar in E13 and adult. Our array analysis also shows that at E13 there is relatively more expression of *Krt5* in the duct (fig. 1c), whereas *Krt14* is more broadly expressed and could label a basal cell population in both the duct and end bud area.

The p63 protein is an additional stem cell marker, and $p63^{-/-}$ mice lack ectodermal tissues [80]. There are two p63 isoforms and a correct balance between TA and ΔNp63 is required for tissue development [81]. ΔNp63 is highly expressed by stem cells and is related to other basal epithelial cells [80]. It is present in basal prostate cells [82], limbal cells, myoepithelial cells, and its expression is also required for the maintenance and proliferation of thymus and basal epidermis stem cells [83]. Furthermore, loss of p63 prevented the full commitment of early endodermal cells to the prostate lineage. However, secretory cells are still formed, suggesting p63 is not required for the commitment to this specific cell lineage [84, 85]. The $p63^{-/-}$ mice do not develop SMGs, suggesting that basal progenitor cells are required for proper SMG development. In the developing SMG there is a relative increase in *p63* expression at E15 and it is still detected in the adult SMG. Our array analysis also shows that it is expressed in both the duct and the end buds of the E13 SMG, which might indicate their expression in basal cells of both compartments.

Influence of Mesenchyme-Produced Growth Factors on Progenitor Cell Fate Decisions

Our observation that epithelial cells from E13 SMGs express a variety of stem cell-related TFs and markers further indicates that the bud, duct, and basal cell niches harbor heterogeneous cell subpopulations. This compartmentalization also raises the possibility that different signaling pathways are needed to maintain/specify these cell populations. Removal of the SMG mesenchyme enables us to study the effect of individual growth factors on epithelial cell fate. The culture of isolated SMG epithelium has been extensively used to study growth factor matrix interactions during epithelial morphogenesis. Here, we have included experimental data where we added exogenous growth factors individually to isolated E13 epithelial rudiments cultured in 3D laminin-111 matrix for only 2 h to measure early downstream changes in gene expression of stem/progenitor cell-related genes associated with cell fate decisions.

A short incubation time was chosen to detect early transcriptional changes and to minimize downstream effects of epithelial cell growth as no change in morphogenesis was observed at 2 h. The effect of growth factors was evaluated by qPCR and normalized to control epithelia cultured in serum-free medium alone (DMEM/Ham's F12 + vitamin C + transferrin). The concentrations of growth factors added were similar to those reported in the literature, i.e. HBEGF at 1 ng/ml, TGF-β_1 and TGF-β_2 at 10 ng/ml, FGF2, FGF8b, BMP2, BMP4 at 100 ng/ml, FGF7 at 200 ng/ml, and FGF10 at 400 ng/ml. There are obvious caveats to this approach but it allows us to address whether a growth factor regulates gene expression of stem/progenitor cell markers present in the E13 epithelium.

Different ES-Related Maintenance Genes Are Downregulated by FGFs

In our analysis, ES-related TF Sox2 was downregulated more than twofold by the addition of FGF7, FGF10, and BMP2 (fig. 2). BMP4 also gave identical results to BMP2 (data not shown). *Klf4* expression was also downregulated by FGF7 and FGF10, which are expressed around the end bud region in vivo, suggesting they promote end bud cell fate. The downregulation of *Sox2* by FGF10 has been observed in other organs such as stomach [86], lung [47] and esophagus [49]. Sox2 may be important for the maintenance of SMG duct cells and its expression may be maintained by other growth factors, such as FGF2 and HBEGF.

FGF7 and FGF10 Control Bud-Related Maintenance/Differentiation Genes

The end bud-associated TFs *cMyc*, *Etv4*, *Etv5*, and *Sox9* are all upregulated by FGF10 and/or FGF7 (fig. 2). The mesenchyme produces FGF10 and FGF7 in vivo and they induce SMG end bud cell proliferation, which is associated with activation of MAPK, ERKs, and cMyc. Since cMyc is highly expressed in the E13 end bud, and both FGF10 and FGF7 promote epithelial growth, cMyc may be a proliferative marker and also label self-renewing end bud progenitor cells. Accordingly, BMPs and TGF-βs promote cell cycle exit and differentiation via cMyc [87, 88], and BMP2 and BMP4 (not shown) also decrease cMyc expression in the E13 SMG epithelium. Although EGF increases proto-oncogenes like cMyc in mES cells [89], E13 SMG cells maintain their *cMyc* expression with HBEGF treatment. Interestingly, none of the tested growth factors influenced Sox10 expression, suggesting it is regulated further downstream (i.e. 2 h is too short) or directly controlled by other growth factors.

Basal Cells Are Regulated by TGF-βs

TGF-βs have been proposed to maintain basal stem/progenitor cells in their niche [90]. In our experiments, TGF-β1 (fig. 2) and TGF-β2 (not shown) gave similar results, both increasing *Krt5* and *Krt14* gene expression (fig. 2). Therefore, the TGF-β family may be important for maintaining the basal stem cell niche. On the other hand, BMP2 and BMP4 (not shown) also increased *Krt14* expression. Interestingly, ΔNp63 is reported to regulate expression of both Krt5 and Krt14, however no direct regulation of Δ*Np63* was observed with any growth factor within 2 h. Since ΔNp63 is a basal marker, interaction with the basement membrane, extracellular matrix, or cell adhesion receptors may influence its expression, rather than with a growth factor alone. Our data also suggests that TGF-β signaling and ΔNp63 regulate the basal keratin expression by different mechanisms.

Conclusion

In this review, we demonstrate that TFs associated with epithelial stem/progenitor cells are present in the E13 SMG, and that, for most, their expression is downregulated towards adulthood. This suggests that several TFs associated with stem/progenitor cell maintenance are not needed in the adult gland, or that the stem/progenitor cells in the adult gland are different from those present in embryonic gland. We have further demonstrated that the E13 SMG epithelium end buds and duct contain different cell types based on their transcriptional profiles. The ductal compartment contains more primitive stem/progenitor cell types based on the expression of TFs related to ES cell self-renewal (Sox2 and Klf4) and higher levels of basal cell keratin (Krt5). Whereas the end buds contain stem/progenitor cell types that express TFs related to progenitor cell maintenance and specification (Etv4, Etv5, Sox9, Sox10). Furthermore, we show that different growth factors produced by the SMG mesenchyme in vivo directly influence the cell fate of distinct cell populations by regulating the transcriptional level of TFs and cytokeratins located in the end bud region, duct area, and the basal cell compartment. Understanding the cell lineage of progenitor cells within the salivary glands is important from a clinical perspective where progenitor cells of specific lineages may be more appropriate than pluripotent stem cells for clinical transplantations to regenerate irradiation-damaged salivary glands.

References

1 Grobstein C: Trans-filter induction of tubules in mouse metanephrogenic mesenchyme. Exp Cell Res 1956;10:424–440.

2 Kratochwil K: Organ specificity in mesenchymal induction demonstrated in the embryonic development of the mammary gland of the mouse. Dev Biol 1969;20:46–71.

3 Kusakabe M, Sakakura T, Sano M, Nishizuka Y: A pituitary-salivary mixed gland induced by tissue recombination of embryonic pituitary epithelium and embryonic submandibular gland mesenchyme in mice. Dev Biol 1985;110:382–391.

4 Nogawa H: Determination of the curvature of epithelial cell mass by mesenchyme in branching morphogenesis of mouse salivary gland. J Embryol Exp Morphol 1983;73:221–232.

5 Nogawa H, Mizuno T: Mesenchymal control over elongating and branching morphogenesis in salivary gland development. J Embryol Exp Morphol 1981;66:209–221.

6 Sakakura T, Nishizuka Y, Dawe CJ: Mesenchyme-dependent morphogenesis and epithelium-specific cytodifferentiation in mouse mammary gland. Science 1976;194:1439–1441.

7 Clayton E, Doupe DP, Klein AM, Winton DJ, Simons BD, Jones PH: A single type of progenitor cell maintains normal epidermis. Nature 2007;446: 185–189.

8 Chang WW: Cell population changes during acinus formation in the postnatal rat submandibular gland. Anat Rec 1974;178:187–201.

9 Cutler LS, Chaudhry AP: Cytodifferentiation of the acinar cells of the rat submandibular gland. Dev Biol 1974;41:31–41.

10 Gresik EW: Postnatal developmental changes in submandibular glands of rats and mice. J Histochem Cytochem 1980;28:860–870.

11 Redman RS, Ball WD: Cytodifferentiation of secretory cells in the sublingual gland of the prenatal rat: a histological, histochemical and ultrastructural study. Am J Anat 1978;153:367–389.

12 Redman RS, Sreebny LM: Morphologic and biochemical observations on the development of the rat parotid gland. Dev Biol 1971;25:248–279.

13 Patel VN, Rebustini IT, Hoffman MP: Salivary gland branching morphogenesis. Differentiation 2006;74: 349–364.

14 Jaskoll T, Witcher D, Toreno L, Bringas P, Moon AM, Melnick M: FGF8 dose-dependent regulation of embryonic submandibular salivary gland morphogenesis. Dev Biol 2004;268:457–469.

15 Jaskoll T, Abichaker G, Witcher D, Sala FG, Bellusci S, Hajihosseini MK, Melnick M: FGF10/FGFR2b signaling plays essential roles during in vivo embryonic submandibular salivary gland morphogenesis. BMC Dev Biol 2005;5:11.

16 Jaskoll T, Zhou YM, Chai Y, Makarenkova HP, Collinson JM, West JD, Hajihosseini MK, Lee J, Melnick M: Embryonic submandibular gland morphogenesis: stage-specific protein localization of FGFs, BMPs, Pax6 and Pax9 in normal mice and abnormal SMG phenotypes in FgfR2-IIIc(+/δ), BMP7(–/–) and Pax6(–/–) mice. Cells Tissues Organs 2002;170:83–98.

17 De Moerlooze L, Spencer-Dene B, Revest J, Hajihosseini M, Rosewell I, Dickson C: An important role for the IIIb isoform of fibroblast growth factor receptor 2 (FGFR2) in mesenchymal-epithelial signalling during mouse organogenesis. Development 2000;127:483–492.

18 Yang A, Schweitzer R, Sun D, Kaghad M, Walker N, Bronson RT, Tabin C, Sharpe A, Caput D, Crum C, McKeon F: P63 is essential for regenerative proliferation in limb, craniofacial and epithelial development. Nature 1999;398:714–718.

19 Jaskoll T, Melnick M: Submandibular gland morphogenesis: Stage-specific expression of TGF-α/EGF, IGF, TGF-β, TNF, and IL-6 signal transduction in normal embryonic mice and the phenotypic effects of TGF-β_2, TGF-β_3, and EGF-R null mutations. Anat Rec 1999;256:252–268.

20 Bianco P, Robey PG: Stem cells in tissue engineering. Nature 2001;414:118–121.

21 Ambler CA, Maatta A: Epidermal stem cells: location, potential and contribution to cancer. J Pathol 2009;217:206–216.

22 Yoshida S, Ohbo K, Takakura A, Takebayashi H, Okada T, Abe K, Nabeshima Y: Sgn1, a basic helix-loop-helix transcription factor delineates the salivary gland duct cell lineage in mice. Dev Biol 2001; 240:517–530.

23 Bullard T, Koek L, Roztocil E, Kingsley PD, Mirels L, Ovitt CE: Ascl3 expression marks a progenitor population of both acinar and ductal cells in mouse salivary glands. Dev Biol 2008;320:72–78.

24 Denny PC, Liu P, Denny PA: Evidence of a phenotypically determined ductal cell lineage in mouse salivary glands. Anat Rec 1999;256:84–90.

25 Man YG, Ball WD, Marchetti L, Hand AR: Contributions of intercalated duct cells to the normal parenchyma of submandibular glands of adult rats. Anat Rec 2001;263:202–214.

26 Sabatini LM, Allen-Hoffmann BL, Warner TF, Azen EA: Serial cultivation of epithelial cells from human and macaque salivary glands. In Vitro Cell Dev Biol 1991;27A:939–948.

27 Horie K, Kagami H, Hiramatsu Y, Hata K, Shigetomi T, Ueda M: Selected salivary-gland cell culture and the effects of isoproterenol, vasoactive intestinal polypeptide and substance P. Arch Oral Biol 1996; 41:243–252.

28 Kim YJ, Kwon HJ, Shinozaki N, Hashimoto S, Shimono M, Cho SW, Jung HS: Comparative analysis of ABCG2-expressing and label-retaining cells in mouse submandibular gland. Cell Tissue Res 2008; 334:47–53.

29 Kimoto M, Yura Y, Kishino M, Toyosawa S, Ogawa Y: Label-retaining cells in the rat submandibular gland. J Histochem Cytochem 2008;56:15–24.

30 Hisatomi Y, Okumura K, Nakamura K, Matsumoto S, Satoh A, Nagano K, Yamamoto T, Endo F: Flow cytometric isolation of endodermal progenitors from mouse salivary gland differentiate into hepatic and pancreatic lineages. Hepatology 2004;39:667–675.

31 Okumura K, Nakamura K, Hisatomi Y, Nagano K, Tanaka Y, Terada K, Sugiyama T, Umeyama K, Matsumoto K, Yamamoto T, Endo F: Salivary gland progenitor cells induced by duct ligation differentiate into hepatic and pancreatic lineages. Hepatology 2003;38:104–113.

32 David R, Shai E, Aframian DJ, Palmon A: Isolation and cultivation of integrin $\alpha_6\beta_1$-expressing salivary gland graft cells: a model for use with an artificial salivary gland. Tissue Eng Part A 2008;14:331–337.
33 Matsumoto S, Okumura K, Ogata A, Hisatomi Y, Sato A, Hattori K, Matsumoto M, Kaji Y, Takahashi M, Yamamoto T, Nakamura K, Endo F: Isolation of tissue progenitor cells from duct-ligated salivary glands of swine. Cloning Stem Cells 2007;9:176–190.
34 Sato A, Okumura K, Matsumoto S, Hattori K, Hattori S, Shinohara M, Endo F: Isolation, tissue localization, and cellular characterization of progenitors derived from adult human salivary glands. Cloning Stem Cells 2007;9:191–205.
35 Kishi T, Takao T, Fujita K, Taniguchi H: Clonal proliferation of multipotent stem/progenitor cells in the neonatal and adult salivary glands. Biochem Biophys Res Commun 2006;340:544–552.
36 Gorjup E, Danner S, Rotter N, Habermann J, Brassat U, Brummendorf TH, Wien S, Meyerhans A, Wollenberg B, Kruse C, von Briesen H: Glandular tissue from human pancreas and salivary gland yields similar stem cell populations. Eur J Cell Biol 2009;88:409–421.
37 Lombaert IM: Regeneration of Irradiated Salivary Glands by Stem Cell Therapy. Groningen, Van Denderen, 2008.
38 Lombaert IM, Brunsting JF, Wierenga PK, Faber H, Stokman MA, Kok T, Visser WH, Kampinga HH, de Haan G, Coppes RP: Rescue of salivary gland function after stem cell transplantation in irradiated glands. PLoS One 2008;3:e2063.
39 Takahashi K, Yamanaka S: Induction of pluripotent stem cells from mouse embryonic and adult fibroblast cultures by defined factors. Cell 2006;126:663–676.
40 Sotomayor P, Godoy A, Smith GJ, Huss WJ: Oct4a is expressed by a subpopulation of prostate neuroendocrine cells. Prostate 2009;69:401–410.
41 Wilschut KJ, Jaksani S, Van Den Dolder J, Haagsman HP, Roelen BA: Isolation and characterization of porcine adult muscle-derived progenitor cells. J Cell Biochem 2008;105:1228–1239.
42 Miyagi S, Masui S, Niwa H, Saito T, Shimazaki T, Okano H, Nishimoto M, Muramatsu M, Iwama A, Okuda A: Consequence of the loss of Sox2 in the developing brain of the mouse. FEBS Lett 2008; 582:2811–2815.
43 Suh H, Consiglio A, Ray J, Sawai T, D'Amour KA, Gage FH: In vivo fate analysis reveals the multipotent and self-renewal capacities of $Sox2^+$ neural stem cells in the adult hippocampus. Cell Stem Cell 2007;1:515–528.
44 Taranova OV, Magness ST, Fagan BM, Wu Y, Surzenko N, Hutton SR, Pevny LH: Sox2 is a dose-dependent regulator of retinal neural progenitor competence. Genes Dev 2006;20:1187–1202.
45 Chen J, Gremeaux L, Fu Q, Liekens D, Van Laere S, Vankelecom H: Pituitary progenitor cells tracked down by side population dissection. Stem Cells 2009; 27:1182–1195.
46 Gontan C, de Munck A, Vermeij M, Grosveld F, Tibboel D, Rottier R: Sox2 is important for two crucial processes in lung development: branching morphogenesis and epithelial cell differentiation. Dev Biol 2008;317:296–309.
47 Ishii Y, Rex M, Scotting PJ, Yasugi S: Region-specific expression of chicken Sox2 in the developing gut and lung epithelium: regulation by epithelial-mesenchymal interactions. Dev Dyn 1998;213:464–475.
48 Okubo T, Pevny LH, Hogan BL: Sox2 is required for development of taste bud sensory cells. Genes Dev 2006;20:2654–2659.
49 Que J, Okubo T, Goldenring JR, Nam KT, Kurotani R, Morrisey EE, Taranova O, Pevny LH, Hogan BL: Multiple dose-dependent roles for Sox2 in the patterning and differentiation of anterior foregut endoderm. Development 2007;134:2521–2531.
50 Que J, Luo X, Schwartz RJ, Hogan BL: Multiple roles for Sox2 in the developing and adult mouse trachea. Development 2009;136:1899–1907.
51 Brembeck FH, Rustgi AK: The tissue-dependent keratin 19 gene transcription is regulated by GKLF/KLF4 and Sp1. J Biol Chem 2000;275:28230–28239.
52 Knox SM, Lombaert, IMA, Reed, X, Hoffman MP: Parasympathetic innervation maintains epithelial progenitor cells during salivary organogenesis. 2010 (submitted).
53 Hudson DL, Guy AT, Fry P, O'Hare MJ, Watt FM, Masters JR: Epithelial cell differentiation pathways in the human prostate: identification of intermediate phenotypes by keratin expression. J Histochem Cytochem 2001;49:271–278.
54 Yamanaka S: Strategies and new developments in the generation of patient-specific pluripotent stem cells. Cell Stem Cell 2007;1:39–49.
55 Zhou Q, Law AC, Rajagopal J, Anderson WJ, Gray PA, Melton DA: A multipotent progenitor domain guides pancreatic organogenesis. Dev Cell 2007;13: 103-114.
56 Steinberg Z, Myers C, Heim VM, Lathrop CA, Rebustini IT, Stewart JS, Larsen M, Hoffman MP: FGFR2b signaling regulates ex vivo submandibular gland epithelial cell proliferation and branching morphogenesis. Development 2005;132:1223–1234.
57 Oikawa T, Yamada T: Molecular biology of the Ets family of transcription factors. Gene 2003;303:11–34.

58 Roehl H, Nusslein-Volhard C: Zebrafish PEA3 and ERM are general targets of FGF8 signaling. Curr Biol 2001;11:503–507.

59 Kubo H, Shimizu M, Taya Y, Kawamoto T, Michida M, Kaneko E, Igarashi A, Nishimura M, Segoshi K, Shimazu Y, Tsuji K, Aoba T, Kato Y: Identification of mesenchymal stem cell (MSC)-transcription factors by microarray and knockdown analyses, and signature molecule-marked MSC in bone marrow by immunohistochemistry. Genes Cells 2009;14:407–424.

60 Kobberup S, Nyeng P, Juhl K, Hutton J, Jensen J: Ets-family genes in pancreatic development. Dev Dyn 2007;236:3100–3110.

61 Foster JW, Dominguez-Steglich MA, Guioli S, Kwok C, Weller PA, Stevanovic M, Weissenbach J, Mansour S, Young ID, Goodfellow PN, et al: Campomelic dysplasia and autosomal sex reversal caused by mutations in an SRY-related gene. Nature 1994;372:525–530.

62 Wagner T, Wirth J, Meyer J, Zabel B, Held M, Zimmer J, Pasantes J, Bricarelli FD, Keutel J, Hustert E, et al: Autosomal sex reversal and campomelic dysplasia are caused by mutations in and around the SRY-related gene Sox9. Cell 1994;79:1111–1120.

63 Piper K, Ball SG, Keeling JW, Mansoor S, Wilson DI, Hanley NA: Novel Sox9 expression during human pancreas development correlates to abnormalities in campomelic dysplasia. Mech Dev 2002; 116:223–226.

64 Lefebvre V, Smits P: Transcriptional control of chondrocyte fate and differentiation. Birth Defects Res C Embryo Today 2005;75:200–212.

65 Bastide P, Darido C, Pannequin J, Kist R, Robine S, Marty-Double C, Bibeau F, Scherer G, Joubert D, Hollande F, Blache P, Jay P: Sox9 regulates cell proliferation and is required for Paneth cell differentiation in the intestinal epithelium. J Cell Biol 2007; 178:635–648.

66 Lynn FC, Smith SB, Wilson ME, Yang KY, Nekrep N, German MS: Sox9 coordinates a transcriptional network in pancreatic progenitor cells. Proc Natl Acad Sci USA 2007;104:10500–10505.

67 Seymour PA, Freude KK, Tran MN, Mayes EE, Jensen J, Kist R, Scherer G, Sander M: Sox9 is required for maintenance of the pancreatic progenitor cell pool. Proc Natl Acad Sci USA 2007;104:1865–1870.

68 Christiano AM: Hair follicle epithelial stem cells get their sox on. Cell Stem Cell 2008;3:3–4.

69 Nowak JA, Polak L, Pasolli HA, Fuchs E: Hair follicle stem cells are specified and function in early skin morphogenesis. Cell Stem Cell 2008;3:33–43.

70 Lincoln J, Kist R, Scherer G, Yutzey KE: Sox9 is required for precursor cell expansion and extracellular matrix organization during mouse heart valve development. Dev Biol 2007;305:120–132.

71 Cheung M, Briscoe J: Neural crest development is regulated by the transcription factor Sox9. Development 2003;130:5681–5693.

72 Stolt CC, Lommes P, Sock E, Chaboissier MC, Schedl A, Wegner M: The Sox9 transcription factor determines glial fate choice in the developing spinal cord. Genes Dev 2003;17:1677–1689.

73 McKeown SJ, Lee VM, Bronner-Fraser M, Newgreen DF, Farlie PG: Sox10 overexpression induces neural crest-like cells from all dorsoventral levels of the neural tube but inhibits differentiation. Dev Dyn 2005;233:430–444.

74 Deal KK, Cantrell VA, Chandler RL, Saunders TL, Mortlock DP, Southard-Smith EM: Distant regulatory elements in a Sox10-β GEO BAC transgene are required for expression of Sox10 in the enteric nervous system and other neural crest-derived tissues. Dev Dyn 2006;235:1413–1432.

75 Nonaka D, Chiriboga L, Rubin BP: Sox10: a pan-schwannian and melanocytic marker. Am J Surg Pathol 2008;32:1291–1298.

76 Purkis PE, Steel JB, Mackenzie IC, Nathrath WB, Leigh IM, Lane EB: Antibody markers of basal cells in complex epithelia. J Cell Sci 1990;97:39–50.

77 Stasiak PC, Purkis PE, Leigh IM, Lane EB: Keratin 19: predicted amino acid sequence and broad tissue distribution suggest it evolved from keratinocyte keratins. J Invest Dermatol 1989;92:707–716.

78 Rebustini IT, Patel VN, Stewart JS, Layvey A, Georges-Labouesse E, Miner JH, Hoffman MP: Laminin α_5 is necessary for submandibular gland epithelial morphogenesis and influences FGFR expression through β_1 integrin signaling. Dev Biol 2007;308:15–29.

79 Walker JL, Menko AS, Khalil S, Rebustini I, Hoffman MP, Kreidberg JA, Kukuruzinska MA: Diverse roles of E-cadherin in the morphogenesis of the submandibular gland: insights into the formation of acinar and ductal structures. Dev Dyn 2008;237:3128–3141.

80 Mills AA, Zheng B, Wang XJ, Vogel H, Roop DR, Bradley A: p63 is a p53 homologue required for limb and epidermal morphogenesis. Nature 1999; 398:708–713.

81 Pellegrini G, Dellambra E, Golisano O, Martinelli E, Fantozzi I, Bondanza S, Ponzin D, McKeon F, De Luca M: p63 identifies keratinocyte stem cells. Proc Natl Acad Sci USA 2001;98:3156–3161.

82 Signoretti S, Waltregny D, Dilks J, Isaac B, Lin D, Garraway L, Yang A, Montironi R, McKeon F, Loda M: p63 is a prostate basal cell marker and is required for prostate development. Am J Pathol 2000;157:1769–1775.
83 Senoo M, Pinto F, Crum CP, McKeon F: p63 is essential for the proliferative potential of stem cells in stratified epithelia. Cell 2007;129:523–536.
84 Kurita T, Medina RT, Mills AA, Cunha GR: Role of p63 and basal cells in the prostate. Development 2004;131:4955–4964.
85 Signoretti S, Pires MM, Lindauer M, Horner JW, Grisanzio C, Dhar S, Majumder P, McKeon F, Kantoff PW, Sellers WR, Loda M: p63 regulates commitment to the prostate cell lineage. Proc Natl Acad Sci USA 2005;102:11355–11360.
86 Nyeng P, Norgaard GA, Kobberup S, Jensen J: FGF10 signaling controls stomach morphogenesis. Dev Biol 2007;303:295–310.
87 Fantozzi I, Platoshyn O, Wong AH, Zhang S, Remillard CV, Furtado MR, Petrauskene OV, Yuan JX: Bone morphogenetic protein-2 upregulates expression and function of voltage-gated K^+ channels in human pulmonary artery smooth muscle cells. Am J Physiol 2006;291:L993–L1004.
88 Pietenpol JA, Stein RW, Moran E, Yaciuk P, Schlegel R, Lyons RM, Pittelkow MR, Munger K, Howley PM, Moses HL: TGF-β_1 inhibition of c-myc transcription and growth in keratinocytes is abrogated by viral transforming proteins with pRB binding domains. Cell 1990;61:777–785.
89 Park JH, Han HJ: Caveolin-1 plays important role in EGF-induced migration and proliferation of mouse embryonic stem cells: Involvement of PI3K/Akt and ERK. Am J Physiol 2009;297:C935–C944.
90 Yamazaki S, Iwama A, Takayanagi S, Eto K, Ema H, Nakauchi H: TGF-β as a candidate bone marrow niche signal to induce hematopoietic stem cell hibernation. Blood 2009;113:1250–1256.

Matthew P. Hoffman
Rm 430, Bldg 30, 30 Convent Dr MSC 43780
National Institute of Dental and Craniofacial Research
National Institutes of Health, Bethesda, MD 20892 (USA)
Tel. +1 301 496 1660, Fax +1 301 402 0897, E-Mail mhoffman@mail.nih.gov

Tucker AS, Miletich I (eds): Salivary Glands. Development, Adaptations and Disease.
Front Oral Biol. Basel, Karger, 2010, vol 14, pp 107–128

Salivary Gland Regeneration

Guy H. Carpenter · Emanuele Cotroneo

Salivary Research Unit, King's College London Dental Institute, London, UK

Abstract

The ability of animal salivary glands to recover from an experimentally-induced atrophic state offers hope that human salivary glands may be regenerated following injury. Examination of the mechanisms of regeneration in animal models has revealed processes which resemble the embryonic formation of salivary glands. Secretory proteins present in regenerated acinar and ductal cells are the same as found in the perinatal salivary glands. The use of microarrays to reveal global gene changes has, in combination with bioinformatic techniques, identified some of the important signalling cascades operating in the early stages of glandular regeneration. The role of stem cells is also considered and would fit in with current ideas of glandular regeneration, however the isolation and subsequent differentiation of stem cells into a normal reflexly secreting gland still requires considerable research.

Clinical Need

Dry mouth, although not a life-threatening condition, greatly reduces the quality of life of patients and severely affects their normal day-to-day experiences. Tasting, eating, speaking and swallowing are all affected, but rampant caries and/or erosion of teeth can leave patients with little confidence to face others in social situations leading to increased isolation. Chronic dry mouth (xerostomia) is measured as a reduction in resting whole-mouth saliva secretion although it should be noted that xerostomia has not been proven to be an objective indicator of glandular hypofunction [1]. The causes of dry mouth may be related either directly to disease, for example by the autoimmune destruction of salivary glands as in Sjögren's syndrome, or indirectly by irradiation treatment of head and neck cancers or by prescribed drugs inducing oral dryness. The last of these is probably the most common cause of dry mouth since over half of all drugs are known to cause dry mouth [2] and the greater number of drugs taken (polypharmacy) greatly increases the chances of developing a dry mouth. In some cases, drug-induced dry mouth can be reversed by the prescribing of an alternative drug that does not have xerostomic side effects. Most common medications cause a subjective feeling of dry mouth by acting centrally on brain centres to reduce fluid secretion, although some have more direct actions due to anticholinergic effects (i.e. antidepressants, antihistamines and antihypertensives); binding to the muscarinic receptors on salivary cells. A very useful website [www.drymouth.info] is available to help clinicians choose better medications without xerostomic side effects. However, for other causes of dry mouth (irradiation and disease-related) there are no recognised treatments to reverse the glandular destruction. Although therapies are available to treat the symptoms of dry mouth, they are not without problems. Frequent sips of water can cause drying of the mouth as water does not have the strong viscoelastic properties of saliva [3] that allow it to wet and stick to oral mucosal (and hard) tissues and yet flow to allow constant replenishment of salivary factors that lubricate and maintain normal oral homeostasis. Even salivary substitutes are poor mimics for saliva as although they may be viscous (by the addition of carboxymethyl cellulose) they do not have similar surface rheology [4] and consequently feels 'sticky' or 'thick' to the patient. Pilocarpine is a parasympathomimetic drug that mimics the main neural stimulus (acetylcholine) for fluid secretion. However, tablet ingestion of this drug leads to excessive sweating and a constant need of urination since they are also mediated by muscarinic receptors. An interesting idea to avoid these problems is the topical application onto the oral mucosa of a choline esterase inhibitor which inhibits the normal breakdown of the parasympathetic neurotransmitter acetylcholine [5], thus prolonging the action of acetylcholine on salivary cells and increasing salivary flow but preventing whole-body side effects.

Salivary gland impairment and atrophy, as a consequence of diseases and medical treatments, leads to reduced saliva secretion. In Sjögren's syndrome, decreased production of saliva has been reported to be caused by the binding of autoantibody to, and inhibition of, muscarinic receptors [6] and later by focal periductal mononuclear cell infiltrates the destruction of salivary cells [2]. Glandular inflammation is usually associated with hypofunction and often considered as causing salivary hypofunction, however recent studies suggest the link is not so obvious. For example, in a short-term duct ligation in the rat in which there was extensive inflammation and secretory hypofunction [7], dexamethasone treatment, which eliminated inflammation, did not affect the salivary hypofunction [8]. In other animal models of lipopolysaccharide-induced inflammation, nitric oxide appears to be an inhibitor of salivary secretion [9] which may also be applicable to Sjögren's syndrome [10]. Acute effects of exposure to X-ray radiation, a major treatment administered for neck and head cancer, also causes irreversible hypofunction of salivary glands [1] which may be due to nitric oxide [11]. Longer term atrophy of irradiated salivary glands appears to be due to loss of salivary gland progenitor/stem cells [12]. Clearly then there is a broad number of reasons why salivary glands may become hypofunctional – either direct inhibition of salivary cells or loss of functional acinar cells.

Salivary Glands Are Highly Adaptable

Salivary glands are highly adapted to each species and show some of the greatest variations of any organ in the body [13]. As such they have evolved to become highly specialized to a particular diet responding to differences in hardness or calorific value of food. This is particularly reflected in secretory granules and their resulting salivary protein composition. In addition to the long-term evolution of salivary glands, they can also adapt acutely to the hardness of food pellets or starvation. Changing diet from the normal hard chow to a softer version but of similar calorific value causes an extensive atrophy of rat parotid glands [14, 15]. Changing the feed back to the normal, harder chow was accompanied by the return of the normal gland weight. These changes in gland size reflect changes in the neural input to the glands from mechanoreceptors located in the periodontal ligament or the gingiva [16]. Enlargement of normal glands can be caused in vivo by the chronic administration of isoprenaline, a β-adrenergic agonist (mimicking sympathetic neural activity). These studies show that salivary glands are inherently adaptable to environmental stimuli and thus have the potential to regenerate following damage to the glands. Salivary glands are densely innervated (fig. 1) by both parasympathetic (causing fluid secretion) and sympathetic (causing most, but not all, protein secretion) nerves which act together rather than antagonistically as in the rest of the body. The strong influence of nerves on salivary gland structure and function are well reviewed [17–19] and nerves should play an important part of salivary gland regeneration. Since salivary secretion is a reflex arc involving taste and chewing receptors [20], regeneration of salivary glands should incorporate the reintegration of parasympathetic nerves to ensure that salivary secretion is upregulated during eating and downregulated during sleep as occurs normally [21].

Animal Models of Glandular Atrophy and Regeneration

Ligation and subsequent deligation of the main excretory duct of salivary glands is remarkable because the gland fully recovers from an atrophic state. Duct ligation-induced atrophy has been known for decades [22, 23] and many cellular changes have been noted (for a summary, see table 1). The most obvious feature of ligation is the rapid loss of differentiated cell types (fig. 2). Acini and granular ducts are no longer apparent as secretory granules, indicated by Alcian blue/periodic acid-Schiff's stain in figure 2, and are autophagocytosed. Invaginations of the plasma membrane caused by the abundant mitochondria, that are characteristic of striated ducts, are also lost. Glandular weight is reduced mainly due to loss of acinar cells following apoptosis [24, 25] but is offset to some extent by proliferation of undifferentiated ductal cells [24, 26]. Removal of ligation allows regeneration of secretory tissue and eventually the glands secrete normal amounts of saliva with broadly normal content of ions and proteins [7, 26, 27]. One of the reasons behind the recovered secretory ability is due to the

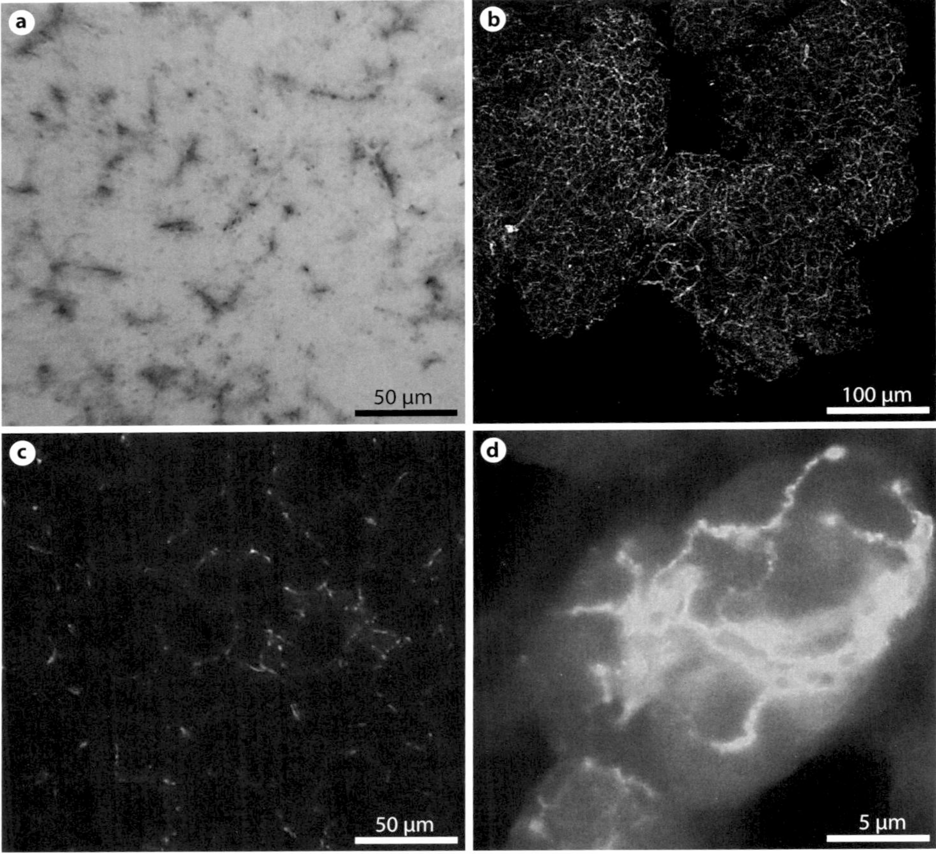

Fig. 1. a–d Nerves in rat salivary glands. Cholinesterase staining of parasympathetic nerves (**a**) and catecholamine staining (**c**) of sympathetic nerves on tissue sections of submandibular glands. Anti-PGP staining (**b**) indicates the extensive network of nerves in larger clumps of acini. Catecholamine staining (**d**) of collagenase-digested acinar cells from adult (unoperated) rat submandibular gland shows individual sympathetic nerve fibres on acinar cells.

re-attachment of the parasympathetic nerves to the target cells, as shown by normal secretion in response to autonomimetic and direct nerve (parasympathetic) stimulation [28]. Confirming that a regenerated gland can secrete reflexly (i.e. in response to taste or chewing) is an important requirement for salivary function.

The ability of the ligation/deligation model to completely regenerate is in contrast to either the partial extirpation of the gland [29] or the irradiated gland model (at higher doses) which do not recover nearly as much [1]. The lack of recovery of the irradiated gland has been attributed to loss of glandular stem cells [30, 31]. Its interesting to note that the irradiated gland looks remarkably similar to the ligated gland (i.e. loss of acini and extended undifferentiated ducts; compare figures in references 32 with 33). This suggests that the irradiated gland might go through a similar

Table 1. Summary of the effects of duct ligation and deligation on salivary glands

	Ligation	Ref.	Deligation	Ref.
Stroma	Enlargement of the intercellular space	24	Reduction of the intercellular space	26, 27, 54
	Inflammatory cell infiltration	106	Decrease of inflammation	
	Denser connective tissue		Recovery of normal gland cytology	
Acini	Apoptosis	24, 25	Proliferation of residual and newly formed acini	54, 107, 108
	Shrunken residual acini Loss of granules	109, 110	Increase of the proportional volume	26
	Loss of mitochondria	111	New acini differentiate from ductal cells	108, 112
Ducts	Luminal dilation due to degranulation	106	Decrease in proliferation activity	54
	Mitotic figures (intercalated and striated ducts)	24		
	Loss of mitochondria	111	Higher than normal duct-to-acinar ratio	26
	Increase of the proportional volume	26		
Myoepithelial cells	Changes in shape	113	Resumption of normal position	114, 115
	Proliferative activity	116, 117		
	Apoptosis	117	Proliferation	115
	Changed position	118		
Gland functionality	Smaller amount of saliva, rich in Na^+	119	Normal saliva production and proteins	7, 27

atrophic process as the ligated gland. Certainly, ligation-induced atrophy is faster acting than a disuse atrophy seen when the parasympathetic nerves are cut [34, 35]. Perhaps the reason that partial extirpation of glands does not cause much glandular regeneration is because the damaged gland does not go through an atrophic process. If so, then irradiated glands (which appear atrophic) have a potential to regenerate by replacing the lost stem cells. This idea is currently being explored by Coppes and colleagues (University of Groningen, the Netherlands) who have used bone marrow-

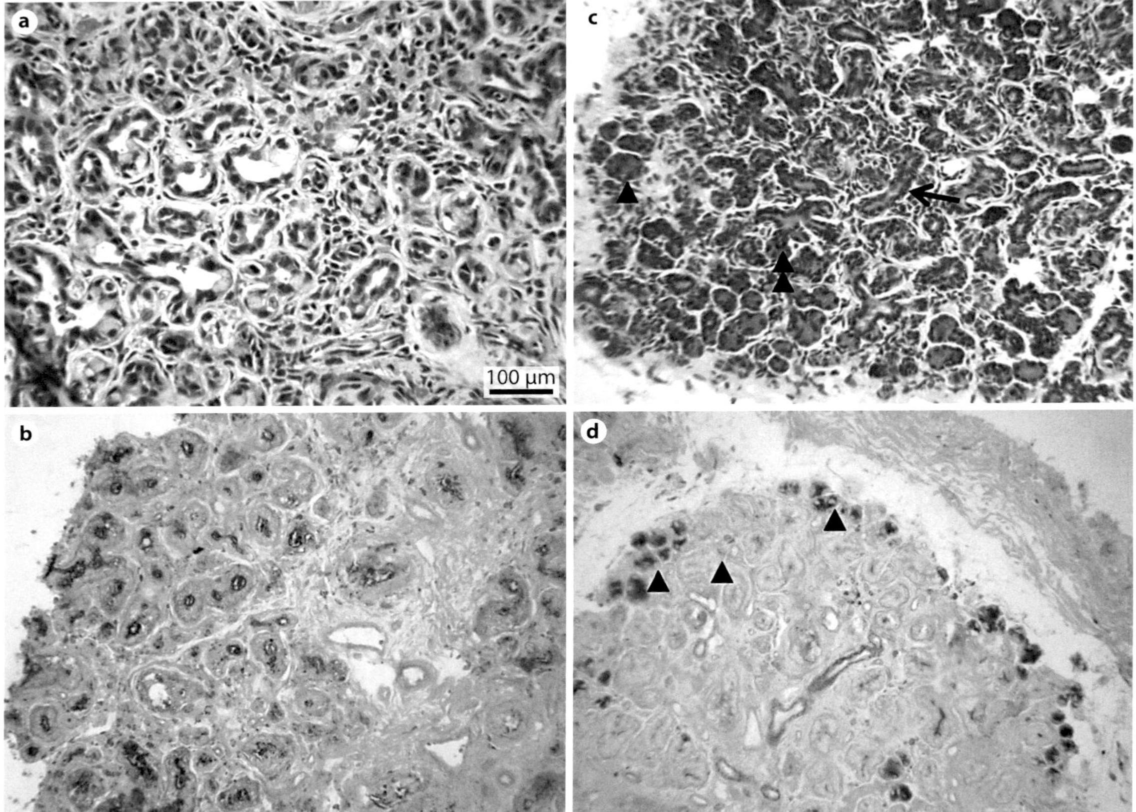

Fig. 2. Histological comparison between ligated (**a**, **b**) and 3-day deligated (regenerated) submandibular glands (**c**, **d**). **a**, **b** Atrophic gland HE and AB/PAS, respectively. HE (**a**) showing luminal dilatation of the duct (arrow), absence of acini and extensive inflammation. AB/PAS (**b**) showing loss of cellular secretory granules and material in the lumena of the ducts (arrow). **c**, **d** 3-Day regenerated gland HE and AB/PAS, respectively. HE (**c**): some acini (arrowhead) and ductal cells (arrow) have recovered some of their size. Acinar-ductal branched structures are often visible (double arrow). AB/PAS (**d**) showing some acini that have recovered their glycoprotein content (arrowhead).

derived mesenchymal stem cells and epithelial stem cell factors to stimulate irradiated rat submandibular gland regeneration [32, 36].

Mechanisms of Maintaining Gland Size

In the mature animal fed a consistent diet, salivary gland size increases proportionally to total animal weight. The normal mechanisms for maintaining consistent gland size is a balance between new cell formation and cell death [37]. In adult glands, levels of mitosis, which can occur in all salivary cell types [38], are relatively low.

Parasympathetic nerve stimulation is known to upregulate mitosis [39] and thus chewing maintains a sufficient level of mitosis to maintain gland size.

Early radiolabelling studies suggested that salivary glands have one source of undifferentiated progenitor cells located in the intercalated ducts that 'streamed' into the different cell types [40]. Further radiolabelling studies subsequently revealed that although the intercalated ductal cells had the highest turnover rates all cells were capable of dividing and maintaining cell proportions [41]. Subsequent analyses using Ki67 [38] or Cre-recombinase studies [42] confirm that normal cell turnover is maintained by more than one population of cells.

The origin of new cells in glands undergoing regeneration following injury has been studied in the ligation/deligation [43–45] and partial extirpation of the rat submandibular gland [29]. Cell labelling by ^{3}H-thymidine, BrdU incorporation and Ki67 staining revealed that new cell formation occurs in both existing cell populations and via budding of ductal structures to form embryonic-like branched structures [29, 46]. The proportion of new cells dividing from existing cells or budding from ducts probably depends on the length of ligation. In 1-week ligated parotid glands a near complete recovery in weight was achieved in only 7 days [47, 48]. Longer ligation periods appears to require longer regeneration periods in a study on rat submandibular glands [33]. Whether the formation of new cells is by division of 'acinar' cells still remaining in the ligated gland or via budding from the ducts is addressed in the next section by analysing the main secretory proteins of each cell type.

Evidence for Regeneration following an Embryonic-Like Pathway

In the first few days of deligation following ligation-induced atrophy, embryonic-like structures appear. Apart from their similar appearance to embryonic structures (fig. 3), branched structures seen in regenerating glands also express some of the same secretory proteins. The submandibular gland proteins A, B, C and D were first identified in developing submandibular glands as being the major secretory proteins [for a review of these proteins, see 49, 50]. Our initial analysis of salivas from regenerated glands revealed that all these proteins, as well as normal adult salivary proteins, were present (fig. 4) in glands following the ligation/deligation procedure. The rapid recovery of mucin production – a protein more associated with adult rather than developing salivary glands [46] – probably reflects the contribution of the existing acinar cell population. Closer analysis of the glands localized one of these proteins SMGB to developing acini [46, and Cotroneo et al., submitted] initially occurring at the extremities of glandular section (fig. 2). Subsequently (day 5 after deligation) ductal branched structures were strongly positive for SMGB. It is these structures that resemble the structures forming during the branching morphogenesis since SMGB is exclusively expressed in the acinar cell precursors in the embryonic gland but not in the adult gland. Occasionally these branched structures can be seen budding off from existing ducts (fig. 5).

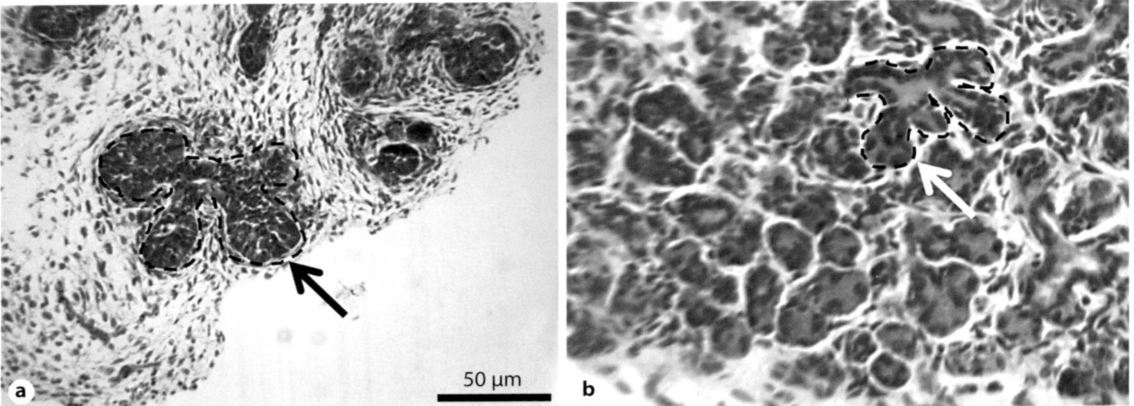

Fig. 3. **a** Branched structures (dotted outline and arrow) occurring in the embryonic submandibular gland (day 19). **b** Branched structures in the 3-day regenerated adult submandibular gland. HE.

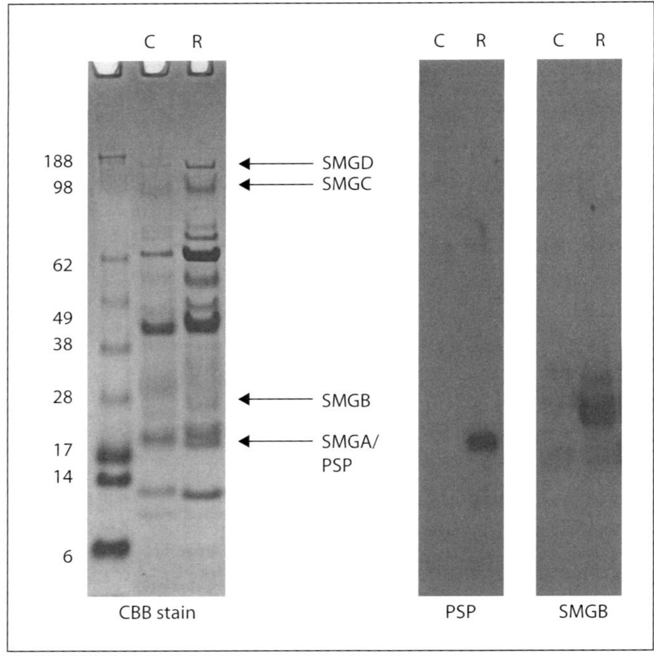

Fig. 4. Salivas from regenerated glands. Embryonic secretory proteins occurring in methacholine-evoked salivas from adult rats before (C = control) and after ligation for 4 weeks and deligation for 8 weeks (R = regenerated). Coomassie staining (CBB stain) indicates several extra proteins including SMG A, B, C & D. Anti-PSP (equivalent to SMGA) and anti-SMGB antibodies have detected those proteins in the salivas from regenerated glands. Position of molecular weight standards (kDa) are indicated.

Similarities between adult salivary gland regeneration and embryogenesis also occur with other proteins (a detailed list of proteins co-expressed in developing glands and regenerating glands, but not normal adult glands is presented in table 2) including the extracellular matrix. The extracellular matrix is a protein layer surrounding cells which collects and concentrates growth factors and is crucial in

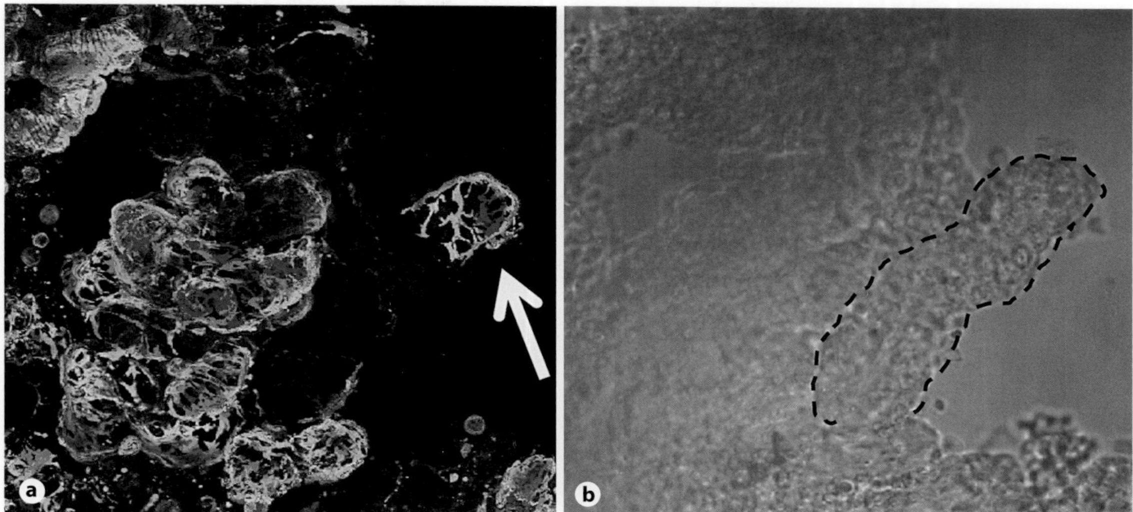

Fig. 5. Anti-smooth muscle actin staining of collagenase cell clumps from a 3-day deligated gland following 2 weeks of ligation. **a** A pseudo-coloured 3D reconstruction of images from a Z-stack series taken by confocal microscopy. Colours reflect depth of field and intensity of staining. Arrow points to budding from a duct to form an embryonic-like branched structure. **b** The duct (dotted outline in phase contrast image) mostly negative for smooth muscle actin.

Table 2. Genes upregulated during development and regeneration but not expressed by normal adult glands in rat [from Cotroneo and Carpenter, unpubl. results]

Gene name	Symbol	Fold change	Expression	Function
Nidogen (entactin)	Nid-1	2.041	Embryonic salivary gland (e14) [53]	Linker of basement membrane components [52]
Cyclin D2	Ccnd2	2.561	Highly expressed in embryonic salivary gland (e14) [120]	Cell-cycle regulator [121]
Sp4 transcription factor	Sp4	2.504	Highly expressed in perinatal salivary gland (p5) [120]	Transcription factor [122]
Secreted frizzled-related protein 1	Sfrp1	2.858	Embryonic salivary gland (e15.5) [123]	Antagonist of the Wnt pathway [124]
Ubiquitin-like, containing PHD and RING finger domains, 1	Uhrf1	9.812	Embryonic and regenerating liver [125]	Cell-cycle regulator [126]
Insulin-like growth (IGF) factor binding protein 5	Igfbp5	3.11	Highly expressed after birth (p1) [120]	Modulate IGFs function [127]

developing glands for controlling branching morphogenesis (see other chapters in this book). In regenerating glands, no obvious changes in laminin, a vital component of the extracellular matrix [51], were found – but levels of nidogen-1 were altered [Cotroneo, unpubl. data]. Nidogen, an extracellular matrix component normally absent in the adult submandibular gland but present in embryonic salivary glands, was significantly upregulated. The presence of an embryonic-specific molecule in the extracellular matrix of adult regenerating gland may promote formation of the ductal branched structures, as occurs during embryonic branching morphogenesis [52, 53].

By studying the expression of novel proteins expressed in deligated adult glands it seems likely that regeneration of salivary glands, at least in the ligation/deligation model, occurs via an embryonic pathway as suggested in early studies by Tamarin [23]. It is worth remembering, however, that self-proliferation of the residual acinar cells (which persisted during atrophy) will also contribute to the repopulation of the gland in the early stage of regeneration [54]. If regeneration does follow an embryonic-like process, then many possibilities for future research become apparent. Branching morphogenesis of developing salivary glands (as reviewed by others in this book) has well-defined extracellular cell matrix requirements [55, 56] and signalling molecules that induce and define branching morphogenesis [57]. These signalling molecules include bone morphogenetic proteins and fibroblast growth factors (FGFs). The introduction of basic FGF into an injured (ligation/deligation [58] or irradiation [59]) gland is one successful treatment which has increased salivary gland regeneration presumably because it aided embryonic-like branching morphogenesis.

The Bioinformatics Approach to Studying Signalling Pathways in Regenerating Glands

Whilst it is crucial to isolate specific molecules involved in glandular regeneration, it is equally important to understand in which signalling cascade such molecules take part in order to identify molecular pathways active during regeneration. In a recent study by the authors, the gene expression profile of rat submandibular gland at two regeneration time-points (3 and 5 days) following 2 weeks of ligation was investigated using microarray technology. In order to identify genes whose expression was changed between the glandular stages, the 3-day regenerating gland was compared to the ligated (atrophic) gland, whilst the 5-day deligated (regenerating) gland was compared to the 3-day samples. The resulting data were then analysed with bioinformatic software which built a list of active signalling pathways. Among these, the mitogen-activated protein kinase (MAPK), the Notch and the Wnt pathway were identified in both the 3- and 5-day regenerating gland (fig. 6).

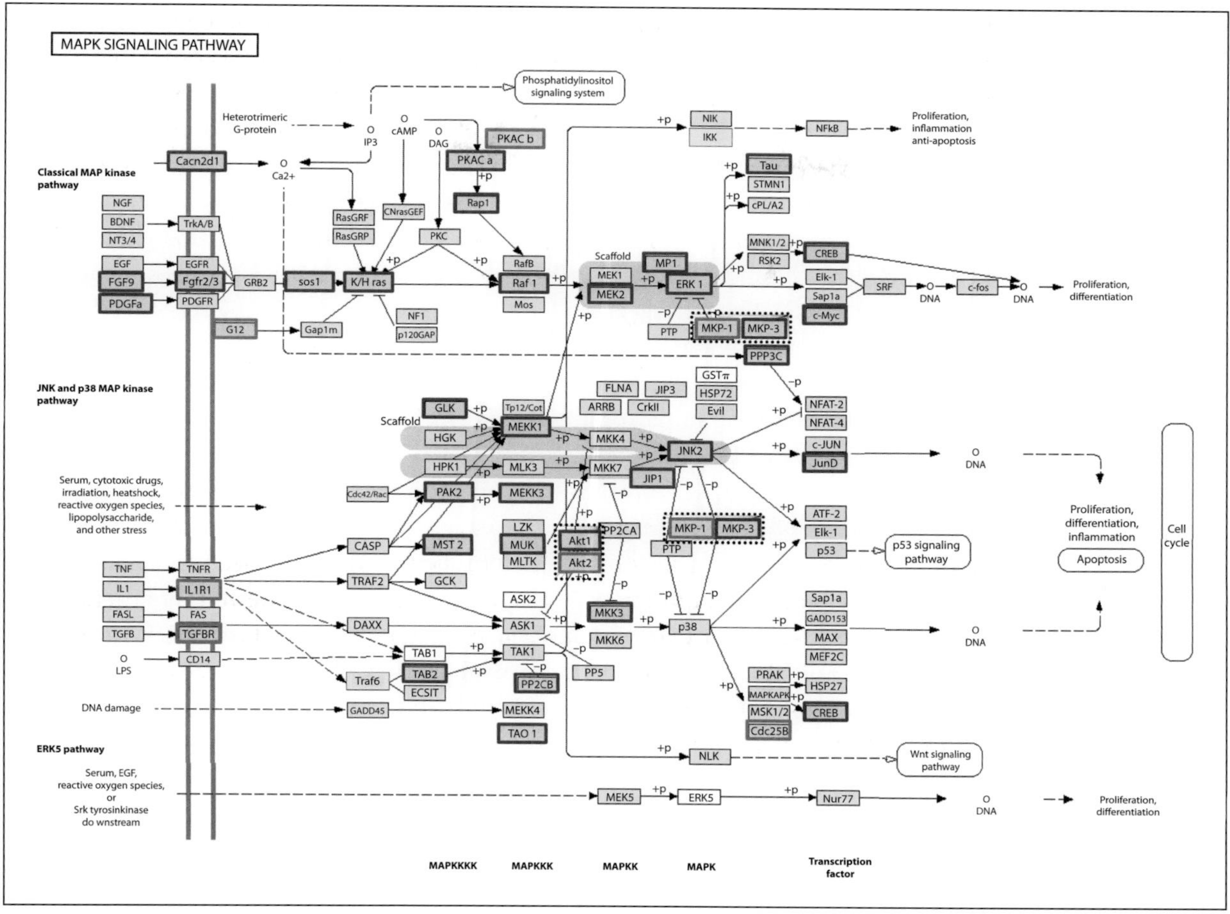

Fig. 6. Graphical representation of MAPK signalling pathway. The genes in bordered boxes represent the genes that changed their expression in the 5-day compared to 3-day regenerated glands (n = 3). The lines terminating with arrows indicate activation, whilst the lines terminating with a straight vertical line indicate inhibition. Some genes were represented in the diagram only with the general gene family name (i.e. MKP, FGFR, FGF, Ras, Sos and Akt). In this cases the specific isoform or the specific family member found in the microarray list was specified.

MAPKs Signalling Pathway

Analysis of the MAPKs signalling revealed a striking difference in the components and the general degree of activation of this pathway between 3 and 5 days of regeneration. In the 3-day deligated compared to the ligated gland, few genes appeared to change; among those downregulated was p38δ, which is commonly activated by cellular stress and pro-inflammatory cytokines and can lead to apoptosis [60]. In contrast, the 5-day regenerating glands showed upregulation of several genes mainly involved in the ERK1/2 and

JNK signalling cascades. The former signalling pathway is usually associated with cell proliferation and survival, whilst the second may lead to either proliferation or apoptosis [61]. A recent study on ligated/deligated rat parotid glands shows similar expression pattern for ERK1/2 and p38 but using more conventional immunohistochemical and biochemical techniques [62]. This study also showed an apparent decrease of activated p38 alongside an increase of activated ERK1/2 in the early stage of regeneration. Due to its role in the stress response, downregulation of p38δ in the 3-day regenerating glands may reflect the decreased inflammation seen in early regeneration [63]. Upregulation of the genes involved in ERK1 signalling cascade, seen at 5 days, may also be responsible for the proliferation and differentiation of the acinar cell precursors at the end of the branching structures observed at this time-point. ERK1/2 signalling activation is mediated by growth factors binding to receptors such as FGF receptors (FGFRs) [64]. Our microarray results indicated FGF9 and its receptor FGFR-2 were upregulated at 5 days. FGF9 has high affinity for FGFR-2 and -3 [65] and thus could be responsible for the downstream activation of the ERK1 signalling cascade eventually leading to proliferation. The importance of FGFR2 in cell proliferation during branching morphogenesis has been derived from an in vitro study on embryonic submandibular gland [66]. This study has proved that FGFR2b is able to promote epithelial proliferation upon FGF7 and FGF10 binding via a MEK1/2-mediated mechanism (MEK2 was also found upregulated in the current study). Although the previous study showed FGFR2 activation upon binding of FGFs different from FGF9, an interesting relationship exists between these ligands. FGF7/10 and FGF9 belong to distinct FGF subfamilies and exert their functions in a reciprocal way, with FGF7/10 signalling from the mesenchyme to the epithelium and FGF9 signalling from the epithelium to the mesenchyme also promoting mesenchymal proliferation [67, 68]. Mesenchymal tissue is absent in the adult submandibular gland, thus it is intriguing as to the source of this signalling in the regenerating gland. Perhaps FGF9 is signalling to some of the mesenchymal derived inflammatory cells within the salivary gland, although more work is required to confirm this.

Notch Signalling Pathway

The fact that the gene expression analysis was carried out on a tissue consisting of a mixed cell population mostly accounts for the complex scenario that emerged in all three signalling pathways discussed here. However, this is particularly true for the Notch signalling cascade, which, according to the bioinformatic analysis, showed activation and deactivation during early submandibular gland regeneration. The apparently contradicting data deriving from the gene expression analysis may reflect the activation in one cell population and the inhibition of Notch signalling in another cell type. Notch signalling commonly implies cell-cell interaction, and it is known to be involved in cell fate decisions via a mechanism called 'lateral inhibition' in which the cell that adopts a fate prevents this choice in the neighbouring cells [69]. During early submandibular gland regeneration (5 days) activation of Notch signalling occurred, mediated by Jagged1 binding to the receptors, as Jagged1 was found to be upregulated along with radical fringe, which has been shown to promote the Jagged1-mediated signalling [70]. A recent study on parotid regeneration (following deligation) also indicated expression of Notch1 and Jagged1 [71].

Wnt Signalling Pathway

Bioinformatic analysis of the microarray data identified upregulation of secreted frizzle-related peptides, inhibitory components of the β-catenin-Wnt (canonical) signalling pathway, in both the time-points (3- and 5-day deligation) of regeneration. These peptides are of interest because of their strong expression during salivary gland organogenesis in the embryo [72]. Further analysis using real-time PCR confirmed upregulation of the Wnt inhibitor secreted frizzle-related peptide 1 (Sfrp1) at 3 days (fig. 7a). Immunocytochemistry of β-catenin showed cell membrane localization (rather than nuclear, if Wnt signalling was active) at all time-points of regeneration, again suggesting the inhibition of this pathway (fig. 7b–d). Axin2, a downstream target of Wnt signalling [73], was also investigated in our study. Levels of Axin2 were not altered in the 3- or 5-day gene lists, thus supporting the initial observations that the Wnt pathway was not active. Pathway analysis of the canonical Wnt signalling at 5 days of regeneration showed the increased presence of more inhibitors of the pathway including Nkd1, duplin and CTpb1. In addition, PP2Ac which has been shown to be involved in stabilizing the β-catenin/E-cadherin complex on the plasma membrane was also upregulated. These results taken together suggested that the canonical Wnt-β-catenin pathway did not play a substantial role in the early submandibular gland regeneration. The increased expression of two downstream target genes of the canonical Wnt pathway, cyclin D2 (table 2) and Myc, presumably reflects induction by the STAT signalling pathway [74].

In the current study the combination of microarray and bioinformatics has been shown to be a valuable procedure to investigate signalling pathways in the regenerating gland. Notably, with regard to MAPK and the Notch pathway, this approach led to similar results obtained with more traditional techniques. However, this procedure presents some limitations, the first of which is related with the microarray analysis itself. The data originating from this analysis requires a subsequent validation step usually by real-time PCR, which, if carefully optimized, should provide a more accurate estimation of gene expression between different experimental conditions. The second limitation of the analysis regards the bioinformatic software, which does not take into consideration cross-talk among molecular pathways, which has to be 'manually' investigated. Considering this, it can be concluded that whilst microarray followed by bioinformatics analysis represents a useful starting point for the analysis of molecular pathways, further validation experiments may be required.

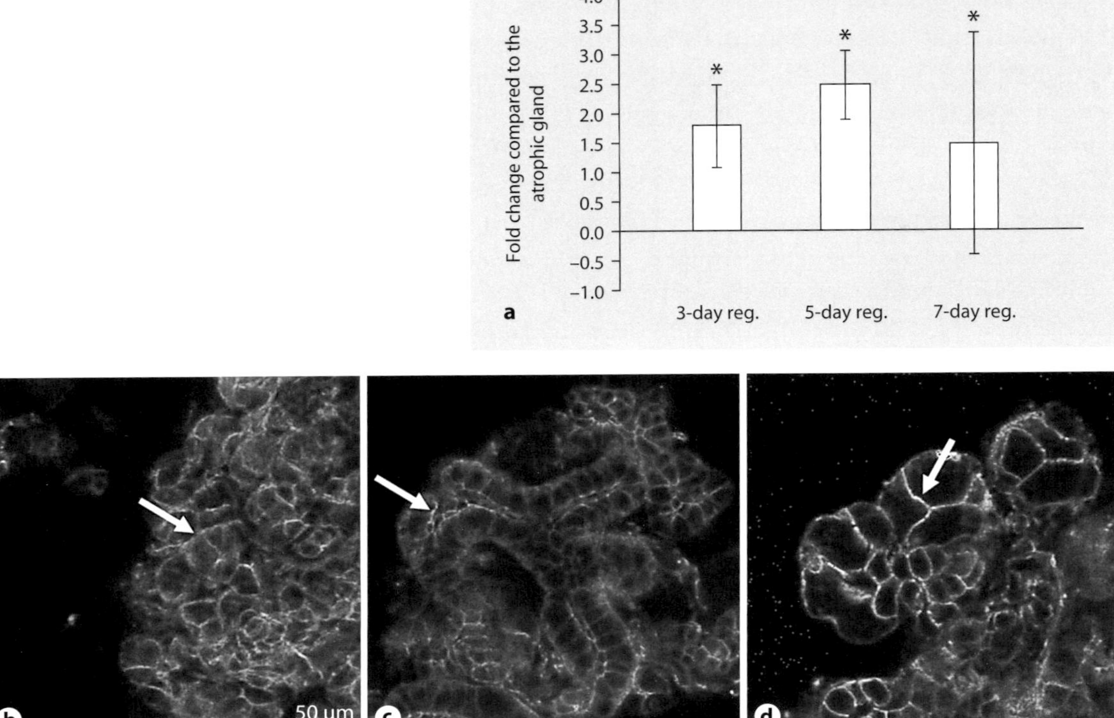

Fig. 7. **a** Real-time PCR analysis showing mean upregulation (* $p < 0.05$) of secreted frizzle-related peptide (sfrp)1 in the 3-, 5- and 7-day regenerated glands (reg.) compared to the atrophic gland (baseline). In the experiment, 5 samples for each condition were run (in duplicate). Bars indicate SEM of the mean. β-Catenin immunofluorescence in 3-, 5- and 7-day regenerated submandibular gland (**b–d** respectively). Collagenase-digested cells were incubated with an anti-β-catenin antibody and viewed by confocal microscopy. At all deligation time-points, β-catenin localized on the cell membrane of both acini and ducts (arrows) where it forms part of the cell-cell junctions rather than in the nucleus, suggesting no activation of the Wnt pathway.

Stem Cell Therapies to Regenerate Salivary Glands

It has been postulated that in addition to self-proliferation of existing cells [75, 76], proliferation and differentiation of progenitor/stem cells also contributes to tissue homeostasis of adult salivary glands in normal and regenerating states [77–79]. Recently some studies, making use of stem/multipotent cell-specific assays and markers, have provided more convincing evidence for the presence of adult salivary gland stem cells in both rodents and humans. In rat adult submandibular gland the presence of a cell population able to form spherical colonies in vitro (indicative of stem cells) and expressing both acinar (AQP5) and ductal (CK19, Na^+,K^+-ATPase) cell markers has been shown [80]. However, the in vivo functional characterization of these cells was not assessed. Okumura et al. [81] have isolated putative salivary stem/progenitor cells from rat salivary gland after duct ligation, which was thought to be necessary to enrich the number of salivary stem/progenitor cells in vivo. The progenitor cells isolated were able to form colonies, were positive for $\alpha_6\beta_1$ integrin and a small number were also positive for c-Kit and Thy-1 (common stem cells markers). Putative human salivary stem cells expressing CD49f$^+$/Thy-1$^+$ have also been identified in the periductal area of submandibular glands [82]. Heat treatment of rats also caused an increase in integrin $\alpha_6\beta_1$-expressing cells from enlarged salivary glands [83]. Once isolated, these cells can differentiate into hepatocyte-like cells but forming salivary-like structures seems elusive. The most convincing in vitro culture of salivary structures used cells isolated from mouse submandibular gland and cultured in an in vitro floating system [84]. In culture these cells showed the ability to form spherical aggregates with proliferation activity (called salispheres). At early culture time-points the salispheres expressed ductal cell markers (CK7 and CK14), and when cultured on a 3D collagen system they started forming ductal structures ending with mucin-positive acini-like formations. Interestingly, the cells within the salisphere were also found positive for stem cell markers such as Sca-1, c-Kit and Musashi-1, which in the adult gland were found to be expressed in the excretory and striated duct cells. Notably, cells isolated from the salispheres showed the ability to restore gland functionality when transplanted in recipient mice subjected to radiation-induced glandular dysfunction. The subset of c-Kit-positive cells seemed to play a crucial role in the glandular restoration. Once the c-Kit-positive fraction was transplanted, the researchers were able to isolate and re-culture this cell subset after long-term transplantation, and eventually to re-transplant them into a secondary recipient (previously irradiated) once more showing amelioration of radiation-induced glandular dysfunction. This is the most convincing evidence yet that stem cells can regenerate salivary glands following injury although it should be noted that the injected cells did not transform into acinar cells but instead appeared to aid regeneration of existing cell populations. Many of the above described studies have identified salivary progenitor/stem cells utilizing stem cell markers identified from other organs. One confounding factor in using c-Kit-positive cells as being progenitor/stem cells is that this receptor is also expressed on other cells, e.g. mast cells [85]. The real test of isolating salivary gland progenitor cells is surely their in vitro and/or in vivo differentiation into salivary glands. Once isolated, progenitor cells should be able to be grown on an extracellular matrix, as shown for embryonic glands [86], to form a branching organ.

Bioengineering of Salivary Glands

There has been a wide variety of strategies to increase salivary (or fluid) flow of either the existing gland [87, 88] or creation of new salivary gland tissue [89], injection of scaffolds that induce branching morphogenesis [90, 91], or the development of an in vitro salivary gland with its subsequent placement into the existing atrophic gland [92, 93]. Transplantation of salivary glands remote from irradiation fields works well [94, 95] despite the potential problems of the gland becoming denervated. Instead, nerves appear to grow in from the surrounding blood vessels which maintains salivary secretion despite an atrophic appearance [96].

Whilst the isolation of cells from normal salivary tissue for growth in vitro has several advantages (compatibility and ease of access), it has a major problem. In culture, salivary cells from tissue explants rapidly dedifferentiate [97]. Cultured cells may express a subset of salivary-specific proteins such as amylase or aquaporin but rarely have the same structured, polarized phenotype found in normal salivary glands. This would be appear to due to the loss of the extracellular influences – in particular, the extracellular matrix and neural stimulation. The first of these have been introduced into cell culture of rat submandibular glands which maintained the normal phenotype for longer [98]. To replicate the input of nerves in vitro is more difficult. The addition of autonomimetics such as isoprenaline, for sympathetic nerves, or carbachol, for parasympathetic nerves, has been used in tissue culture. However, this leads to the continuous stimulation of cells, rather than the intermittent nerve traffic which normally occurs and has been recorded in vivo [99]. Similar concerns in cardiac cell culture have led to the creation of electrical stimulators within culture dishes [100] which could be used to advantage with salivary cultures.

For the regeneration of glands in vivo the extracellular matrix could be replicated/supplemented by injection into the existing gland of chitosan particles which appears to increase branching morphogeneis [91] for embryonic structures. However, the in vivo regeneration of salivary glands may be limited by the regrowth of nerves into damaged or diseased salivary glands. The role of nerves in maintaining normal functional phenotypes in vivo has been well reviewed [35] and peripheral nerves are generally regarded as having limited potential to regenerate. However, in salivary glands there is experimental evidence that nerves can regenerate to innervate salivary glands [28, 96, 101, 102]. Examining the factors that guide nerves towards acini would appear to be an important event for restarting glandular secretion.

There is a similar lack of evidence concerning the angiogenesis of salivary glands and how it may alter during atrophy and regeneration. Salivary glands have a very rich blood supply second only to cardiac muscle [103]. Beautiful demonstrations of blood vessels in salivary glands, by injecting resin into vessels and eroding away the salivary tissue [104], reveal the dense network of vessels in a normal gland. The importance of restoring the blood supply to an irradiated gland [88] have also shown that although blood supply usually meets the demands of normal salivary secretion (in terms of fluid and oxygen delivery), it can become deficient in pathology [105].

Conclusions

The isolation of stem cells that may differentiate into salivary tissue appears to be the most likely route to clinical regeneration of salivary glands, although this would be just the starting point. Presumably stem cells exist in normal glands but the signals that cause their further development into salivary tissue do not appear unless injury occurs. If regeneration does follow an embryonic-like pathway then the extracellular matrix would appear to be an important factor. However, the nature of the injury may also be important. Studies in animal models indicate that ligation-induced atrophy allows glandular regeneration whereas irradiation-induced injury does not, a comparison of these two models may yield some of the important signals that allow regeneration to proceed. Once regeneration starts, the role of nerves in forming a functional gland that reflexly secretes and maintains the highly polarized cell types necessary for salivary secretion must be considered. Transplanted salivary glands have shown that incorrectly innervated salivary glands, growing from nearby blood vessels that respond to changes in blood flow with exercise, for example, may lead to patient dissatisfaction with the regenerated gland.

Acknowledgements

The authors gratefully acknowledge funding by the Wellcome Trust and helpful discussions with Prof. Gordon Proctor and other members of the Salivary Research Unit at King's College London Dental Institute.

References

1 Fox PC: Acquired salivary dysfunction – drugs and radiation. Ann NY Acad Sci 1998;842:132–137.
2 Ship JA, Pillemer SR, Baum BJ: Xerostomia and the geriatric patient. J Am Geriatr Soc 2002;50:535–543.
3 Bongaerts JHH, Rossetti D, Stokes JR: The lubricating properties of human whole saliva. Tribol Lett 2007;27:277–287.
4 Proctor GB, Hamdan S, Carpenter GH, Wilde P: A statherin- and calcium-enriched layer at the air interface of human parotid saliva. Biochem J 2005; 389:111–116.
5 Khosravani N, Ekstrom J, Birkhed D: Intraoral stimulation of salivary secretion with the cholinesterase inhibitor physostigmine as a mouth spray: a pilot study in healthy volunteers. Arch Oral Biol 2007;52:1097–1101.
6 Dawson LJ, Fox PC, Smith PM: Sjögren's syndrome – the non-apoptotic model of glandular hypofunction. Rheumatology 2006;45:792–798.
7 Carpenter GH, Osailan SM, Correia P, Paterson KP, Proctor GB: Rat salivary gland ligation causes reversible secretory hypofunction. Acta Physiol 2007;189:241–249.
8 Correia PN, Carpenter GH, Osailan SM, Paterson KL, Proctor GB: Acute salivary gland hypofunction in the duct ligation model in the absence of inflammation. Oral Dis 2008;14:520–528.
9 Lomniczi A, Mohn C, Faletti A, Franchi A, Rettori V, Elverdin JC: Inhibition of salivary secretion by lipopolysaccharide: possible role of prostaglandins. J Dent Res 2001;80:967.
10 Caulfield VL, Balmer C, Dawson LJ, Smith PM: A role for nitric oxide-mediated glandular hypofunction in a non-apoptotic model for Sjögren's syndrome. Rheumatology 2009;48:727–733.

11 De la Cal C, Lomniczi A, Mohn CE, De Laurentiis A, Casal M, Chiarenza A, Paz D, McCann SM, Rettori V, Elverdin JC: Decrease in salivary secretion by radiation mediated by nitric oxide and prostaglandins. Neuroimmunomodulation 2006;13:19–27.
12 Coppes RP, Zeilstra LJW, Kampinga HH, Konings AWT: Early to late sparing of radiation damage to the parotid gland by adrenergic and muscarinic receptor agonists. Br J Cancer 2001;85:1055–1063.
13 Tandler B, Phillips CJ: Microstructure of mammalian salivary glands and its relationship to diet; in Garrett JR, Ekstrom J, Anderson LC (eds): Glandular Mechanisms of Salivary Secretion. Basel, Karger, 1998, pp 21–35.
14 Hall HD, Schneyer CA: Influence of liquid ration on structure & function of salivary glands of rat. J Dent Res 1964;43:880.
15 Hand AR, Ho B: Liquid-diet-induced alterations of rat parotid acinar cells studied by electron microscopy and enzyme cytochemistry. Arch Oral Biol 1981;26:369–380.
16 Hector MP, Linden RW: Reflexes of salivary secretion; in Garrett JR, Ekstrom J, Anderson LC (eds): Neural Mechanisms of Salivary Secretion. Basel, Karger, 1999, pp 196–217.
17 Garrett JR: The proper role of nerves in salivary secretion – a review. J Dent Res 1987;66:387–397.
18 Proctor GB, Carpenter GH: Regulation of salivary gland function by autonomic nerves. Auton Neurosci 2007;133:3–18.
19 Ekström J: Role of nonadrenergic, noncholinergic autonomic transmitters in salivary glandular activities in vivo; in Garrett JR, Ekstrom J, Anderson LC (eds): Neural Mechanisms of Salivary Secretion. Front Oral Biol. Basel, Karger, 1999, pp 94–130.
20 Matsuo R: Central connections for salivary innervations and efferent impulse formation; in Garrett JR, Ekstrom J, Anderson LC (eds): Neural Mechanisms of Salivary Secretion. Front Oral Biol. Basel, Karger, 1999, pp 26–43.
21 Dawes C: How much saliva is enough for avoidance of xerostomia? Caries Res 2004;38:236–240.
22 Tamarin A: Submaxillary gland recovery from obstruction. 1. Overall changes and electron microscopic alterations of granular duct cells. J Ultrastruct Res 1971;34:276–287.
23 Tamarin A: Submaxillary gland recovery from obstruction. 2. Electron microscopic alterations of acinar cells. J Ultrastruct Res 1971;34:288–297.
24 Walker NI, Gobe GC: Cell death and cell proliferation during atrophy of the rat parotid gland induced by duct obstruction. J Pathol 1987;153:333–344.
25 Takahashi S, Gobe GC, Yoshimura Y, Kohgo T, Yamamoto T, Wakita M: Participation of the Fas and Fas ligand systems in apoptosis during atrophy of the rat submandibular glands. Int J Exp Pathol 2007;88:9–17.
26 Scott J, Liu P, Smith PM: Morphological and functional characteristics of acinar atrophy and recovery in the duct-ligated parotid gland of the rot. J Dent Res 1999;78:1711–1719.
27 Osailan SM, Proctor GB, Carpenter GH, Paterson KL, McGurk M: Recovery of rat submandibular salivary gland function following removal of obstruction: a sialometrical and sialochemical study. Int J Exp Pathol 2006;87:411–423.
28 Carpenter GH, Khosravani N, Ekstrom J, Osailan SM, Paterson KP, Proctor GB: Altered plasticity of the parasympathetic innervation in the recovering rat submandibular gland following extensive atrophy. Exp Physiol 2009;94:213–219.
29 Takahashi S, Wakita M: Regeneration of the intralobular duct and acinus in rat submandibular glands after YAG laser irradiation. Arch Histol Cytol 1993;56:199–206.
30 Coppes RP, Zeilstra LJW, Kampinga HH, Konings AWT: Early to late sparing of radiation damage to the parotid gland by adrenergic and muscarinic receptor agonists. Br J Cancer 2001;85:1055–1063.
31 Konings AWT, Coppes RP, Vissink A: On the mechanism of salivary gland radiosensitivity. Int J Radiat Oncol Biol Phys 2005;62:1187–1194.
32 Lombaert IMA, Brunsting JF, Wierenga PK, Kampinga HH, de Haan G, Coppes RP: Keratinocyte growth factor prevents radiation damage to salivary glands by expansion of the stem/progenitor pool. Stem Cells 2008;26:2595–2601.
33 Osailan SM, Proctor GB, Carpenter GH, Paterson KL, McGurk M: Recovery of rat submandibular salivary gland function following removal of obstruction: a sialometrical and sialochemical study. Int J Exp Pathol 2006;87:411–423.
34 Carpenter GH, Proctor GB, Garrett JR: Preganglionic parasympathectomy decreases salivary SIgA secretion rates from the rat submandibular gland. J Neuroimmunol 2005;160:4–11.
35 Proctor GB: Effects of autonomic denervations on protein secretion and synthesis by salivary glands; in Garrett JR, Ekstrom J, Anderson LC (eds): Neural Mechanisms of Salivary Secretion. Front Oral Biol. Basel, Karger, 1999, pp 150–165.
36 Lombaert IMA, Wierenga PK, Kok T, Kampinga HH, de Haan G, Coppes RP: Mobilization of bone marrow stem cells by granulocyte colony-stimulating factor ameliorates radiation-induced damage to salivary glands. Clin Cancer Res 2006;12:1804–1812.

37 Redman RS: On approaches to the functional restoration of salivary glands damaged by radiation therapy for head and neck cancer, with a review of related aspects of salivary gland morphology and development. Biotech Histochem 2008;83:103–130.
38 Ihrler S, Blasenbreu-Vogt S, Sendelhofert A, Rossle M, Harrison JD, Lohrs U: Regeneration in chronic sialadenitis: an analysis of proliferation and apoptosis based on double immunohistochemical labelling. Virchows Archiv 2004;444:356–361.
39 Schneyer CA, Humphreys-Beyer MG, Hall HD, Jirakulsomchok D: Mitogenic activity of rat salivary glands after electrical stimulation of parasympathetic nerves. Am J Physiol 1993;264:G935–G938.
40 Zajicek G, Yagil C, Michaeli Y: The streaming submandibular gland. Anat Rec 1985;213:150–158.
41 Denny PC, Ball WD, Redman RS: Salivary glands: a paradigm for diversity of gland development. Crit Rev Oral Biol Med 1997;8:51–75.
42 Bullard T, Koek L, Roztocil E, Kingsley PD, Mirels L, Ovitt CE: Ascl3 expression marks a progenitor population of both acinar and ductal cells in mouse salivary glands. Dev Biol 2008;320:72–78.
43 Takahashi S, Nakamura S, Suzuki R, Islam N, Domon T, Yamamoto T, Wakita M: Apoptosis and mitosis of parenchymal cells in the duct-ligated rat submandibular gland. Tissue Cell 2000;32:457–463.
44 Takahashi S, Nakamura S, Domon T, Yamamoto T, Wakita M: Active participation of apoptosis and mitosis in sublingual gland regeneration of the rat following release from duct ligation. J Mol Histol 2005;36:199–205.
45 Takahashi S, Kohgo T, Nakamura S, Arambawatta A, Domon T, Yamamoto T, Wakita M: Biological behavior of myoepithelial cells in the regeneration of rat atrophied sublingual glands following release from duct ligation. J Mol Histol 2005;36:373–379.
46 Cotroneo EM, Proctor GB, Paterson KL, Carpenter GH: Early markers of regeneration following ductal ligation in the rat submandibular gland. Cell Tissue Res 2008;332:227–235.
47 Takahashi S, Schoch E, Walker NI: Origin of acinar cell regeneration after atrophy of the rat parotid induced by duct obstruction. Int J Exp Pathol 1998; 79:293–301.
48 Walker NI, Gobe GC: Cell death and cell proliferation during atrophy of the rat parotid gland induced by duct obstruction. J Pathol 1987;153:333–344.
49 Ball WD, Mirels L, Hand AR: Psp and Smgb: a model for developmental and functional regulation in the rat major salivary glands. Biochem Soc Trans 2003;31:777–780.
50 Mirels L, Miranda AJ, Ball WD: Characterization of the rat salivary gland B1-immunoreactive proteins. Biochem J 1998;330:437–444.
51 Hosokawa Y, Takahashi Y, Kadoya Y, Yamashina S, Nomizu M, Yamada Y, Nogawa H: Significant role of laminin-1 in branching morphogenesis of mouse salivary epithelium cultured in basement membrane matrix. Dev Growth Diff 1999;41:207–216.
52 Ho MSP, Bose K, Mokkapati S, Nischt R, Smyth N: Nidogens – extracellular matrix linker molecules. Microsc Res Tech 2008;71:387–395.
53 Kadoya Y, Salmivirta K, Talts JF, Kadoya K, Mayer U, Timpl R, Ekblom P: Importance of nidogen binding to laminin γ_1 for branching epithelial morphogenesis of the submandibular gland. Development 1997;124:683–691.
54 Takahashi S, Shinzato K, Nakamura S, Domon T, Yamamoto T, Wakita M: Cell death and cell proliferation in the regeneration of atrophied rat submandibular glands after duct ligation. J Oral Pathol Med 2004;33:23–29.
55 Patel VN, Knox SM, Likar KM, Lathrop CA, Hossain R, Eftekhari S, Whitelock JM, Elkin M, Vlodavsky I, Hoffman MP: Heparanase cleavage of perlecan heparan sulfate modulates FGF10 activity during ex vivo submandibular gland branching morphogenesis. Development 2007;134:4177–4186.
56 Rebustini IT, Patel VN, Stewart JS, Layvey A, Georges-Labouesse E, Miner JH, Hoffman MP: Laminin α_5 is necessary for submandibular gland epithelial morphogenesis and influences FGFR expression through β_1 integrin signaling. Dev Biol 2007;308:15–29.
57 Steinberg Z, Myers C, Heim VM, Lathrop CA, Rebustini IT, Stewart JS, Larsen M, Hoffman MP: FGFR2b signaling regulates ex vivo submandibular gland epithelial cell proliferation and branching morphogenesis. Development 2005;132:1223–1234.
58 Okazaki Y, Kagami H, Hattori T, Hishida S, Shigetomi T, Ueda M: Acceleration of rat salivary gland tissue repair by basic fibroblast growth factor. Arch Oral Biol 2000;45:911–919.
59 Thula TT, Schultz G, Tran-Son-Tay R, Batich C: Effects of EGF and bFGF on irradiated parotid glands. Ann Biomed Eng 2005;33:685–695.
60 Hu MC, Wang YP, Mikhail A, Qiu WR, Tan TH: Murine p38-δ mitogen-activated protein kinase, a developmentally regulated protein kinase that is activated by stress and proinflammatory cytokines. J Biol Chem 1999;274:7095–7102.
61 Chang LF, Karin M: Mammalian MAP kinase signalling cascades. Nature 2001;410:37–40.
62 Dang H, Elliott JJ, Lin AL, Zhu B, Katz MS, Yeh CK: Mitogen-activated protein kinase up-regulation and activation during rat parotid gland atrophy and regeneration: role of epidermal growth factor and β_2-adrenergic receptors. Differentiation 2008;76: 546–557.

63 Cotroneo E, Proctor GB, Paterson KL, Carpenter GH: Early markers of regeneration following ductal ligation in rat submandibular gland. Cell Tissue Res 2008;332:227–235.

64 Schlessinger J: Cell signaling by receptor tyrosine kinases. Cell 2000;103:211–225.

65 Hecht D, Zimmerman N, Bedford M, Avivi A, Yayon A: Identification of fibroblast-growth-factor-9 (FGF9) as a high-affinity, heparin dependent ligand for FGF receptor-3 and receptor-2 but not for FGF receptor-1 and receptor-4. Growth Factors 1995;12:223–233.

66 Steinberg Z, Myers C, Heim VM, Lathrop CA, Rebustini IT, Stewart JS, Larsen M, Hoffman MP: FGFR2b signaling regulates ex vivo submandibular gland epithelial cell proliferation and branching morphogenesis. Development 2005;132:1223–1234.

67 Beenken A, Mohammadi M: The FGF family: biology, pathophysiology and therapy. Nat Rev Drug Discov 2009;8:235–253.

68 Geske MJ, Zhang XQ, Patel KK, Ornitz DM, Stappenbeck TS: FGF9 signaling regulates small intestinal elongation and mesenchymal development. Development 2008;135:2959–2968.

69 Ehebauer M, Hayward P, Arias AM: Notch, a universal arbiter of cell fate decisions. Science 2006;314:1414–1415.

70 Yang LT, Nichols JT, Yao C, Manilay JO, Robey EA, Weinmaster G: Fringe glycosyltransferases differentially modulate Notch1 proteolysis induced by Delta1 and Jagged1. Mol Biol Cell 2005;16:927–942.

71 Dang H, Lin AL, Zhang B, Zhang HM, Katz MS, Yeh CK: Role for notch signaling in salivary acinar cell growth and differentiation. Dev Dyn 2009;238:724–731.

72 Leimeister C, Bach A, Gessler M: Developmental expression patterns of mouse sFRP genes encoding members of the secreted frizzled-related protein family. Mech Dev 1998;75:29–42.

73 Jho EH, Zhang T, Domon C, Joo CK, Freund JN, Costantini F: Wnt/β-catenin/Tcf signaling induces the transcription of Axin2, a negative regulator of the signaling pathway. Mol Cell Biol 2002;22:1172–1183.

74 Buettner R, Mora LB, Jove R: Activated STAT signaling in human tumors provides novel molecular targets for therapeutic intervention. Clin Cancer Res 2002;8:945–954.

75 Chai Y, Klauser DK, Denny PA, Denny PC: Proliferative and structural differences between male and female mouse submandibular glands. Anat Rec 1993;235:303–311.

76 Denny PC, Chai Y, Klauser DK, Denny PA: Parenchymal cell proliferation and mechanisms for maintenance of granular duct and acinar cell populations in adult male mouse submandibular gland. Anat Rec 1993;235:475–485.

77 Denny PC, Denny PA: Dynamics of parenchymal cell division, differentiation, and apoptosis in the young adult female mouse submandibular gland. Anat Rec 1999;254:408–417.

78 Man YG, Ball WD, Culp DJ, Hand AR, Moreira JE: Persistence of a perinatal cellular phenotype in submandibular glands of adult rat. J Histochem Cytochem 1995;43:1203–1215.

79 Man YG, Ball WD, Marchetti L, Hand AR: Contributions of intercalated duct cells to the normal parenchyma of submandibular glands of adult rats. Anat Rec 2001;263:202–214.

80 Kishi T, Takao T, Fujita K, Taniguchi H: Clonal proliferation of multipotent stem/progenitor cells in the neonatal and adult salivary glands. Biochem Biophys Res Commun 2006;340:544–552.

81 Okumura K, Nakamura K, Hisatomi Y, Nagano K, Tanaka Y, Terada K, Sugiyama T, Umeyama K, Matsumoto K, Yamamoto T, Endo F: Salivary gland progenitor cells induced by duct ligation differentiate into hepatic and pancreatic lineages. Hepatology 2003;38:104–113.

82 Sato A, Okumura K, Matsumoto S, Hattori K, Hattori S, Shinohara M, Endo F: Isolation, tissue localization, and cellular characterization of progenitors derived from adult human salivary glands. Cloning Stem Cells 2007;9:191–205.

83 David R, Shai E, Aframian DJ, Palmon A: Isolation and cultivation of integrin $\alpha_6\beta_1$-expressing salivary gland graft cells: a model for use with an artificial salivary gland. Tissue Eng Part A 2008;14:331–337.

84 Lombaert IMA, Brunsting JF, Wierenga PK, Faber H, Stockman MA, Kok T, Wisser WH, Kapinga HH, de Haan G, Coppes RP: Rescue of salivary gland function after stem cell transplantation in irradiated glands. PLoS One 2008;3:e2063.

85 Okayama Y, Kawakami T: Development migration, and survival of mast cells. Immunol Res 2006;34:97–115.

86 Wei C, Larsen M, Hoffman MP, Yamada KM: Self-organization and branching morphogenesis of primary salivary epithelial cells. Tissue Eng 2007;13:721–735.

87 Cotrim AP, Baum BJ: Gene therapy: some history, applications, problems, and prospects. Toxicol Pathol 2008;36:97–103.

88 Baum BJ, Zheng C, Cotrim AP, McCullagh L, Goldsmith CM, Brahim JS, Atkinson JC, Turner RJ, Liu S, Nikolov N, Illei GG: Aquaporin-1 gene transfer to correct radiation-induced salivary hypofunction. Handb Exp Pharmacol 2009;403–418.

89 Kagami H, Wang S, Hai B: Restoring the function of salivary glands. Oral Dis 2008;14:15–24.
90 Yang TL, Young TH: The specificity of chitosan in promoting branching morphogenesis of progenitor salivary tissue. Biochem Biophys Res Commun 2009;381:466–470.
91 Yang TL, Young TH: The enhancement of submandibular gland branch formation on chitosan membranes. Biomaterials 2008;29:2501–2508.
92 Joraku A, Sullivan CA, Yoo J, Atala A: In vitro reconstitution of three-dimensional human salivary gland tissue structures. Differentiation 2007;75:318–324.
93 Sugito T, Kagami H, Hata K, Nishiguchi H, Ueda M: Transplantation of cultured salivary gland cells into an atrophic salivary gland. Cell Transplant 2004;13:691–699.
94 Geerling G, Collin JRO, Dart JKG: Ophthalmic experience with submandibular gland transplantation for severe dry eyes. Laryngoscope 2009;119:1445–1446.
95 Li YM, Zhang Y, Shi L, Xiang B, Cong X, Zhang YY, Wu LL, Yu GY: Isoproterenol improves secretion of transplanted submandibular glands. J Dent Res 2009;88:477–482.
96 Gerling G, Garrett JR, Paterson KP, Sieg JP, Collin RO, Carpenter GH, Hakim SG, Lauer I, Proctor GB: Innervation and secretory function of transplanted human submandibular glands. Transplantation 2008;85:135–140.
97 Quissell DO, Redman RS, Mark MR: Short-term primary culture of acinar-intercalated duct complexes from rat submandibular glands. In Vitro Cell Dev Biol 1986;22:469–480.
98 Quissell DO, Redman RS, Barzen KA, McNutt RL: Effects of oxygen, insulin, and glucagon concentrations on rat submandibular acini in serum-free primary culture. In Vitro Cell Dev Biol Anim 1994;30A:833–842.
99 Matsuo R: Electrophysiological analysis of synaptic inputs into the superior salivatory nucleus of rats. J Pharmacol Sci 2003;91:40P.
100 Tandon N, Cannizzaro C, Chao PHG, Maidhof R, Marsano A, Au HTH, Radisic M, Vunjak-Novakovic G: Electrical stimulation systems for cardiac tissue engineering. Nat Protoc 2009;4:155–173.
101 Isomura ET, Yoshitomi K, Hamaguchi M, Kogo M: Saliva secretion stimulated by grafted nerve in submandibular gland allograft in dogs. Transplantation 2007;83:759–763.
102 Ohlin P, Perec C: Secretory responses and choline acetylase of the rat's submaxillary gland after duct ligation. Experimentia 1967;23:248–249.
103 Smaje LH: Capillary dynamics in salivary glands; in Garrett JR, Ekstrom J, Anderson LC (eds): Glandular Mechanisms of Salivary Secretion. Front Oral Biol. Basel, Karger, 1998, pp 118–131.
104 Ohtani O, Ohtsuka A, Lipsett J, Gannon B: The microvasculature of rat salivary glands – a scanning electron-microscopic study. Acta Anat 1983;115:345–356.
105 Berggreen E, Nylokken K, Delaleu N, Hajdaragic-Ibricevic H, Jonsson MV: Impaired vascular responses to parasympathetic nerve stimulation and muscarinic receptor activation in the submandibular gland in non-obese diabetic mice. Arthritis Res Ther 2009;11.
106 Norberg LE, Abok K, Lundquist PG: Effects of ligation and irradiation on the submaxillary glands in rats. Acta Otolargol 1988;105:181–192.
107 Takahashi S, Nakamura S, Domon T, Yamamoto T, Wakita M: Active participation of apoptosis and mitosis in sublingual gland regeneration of the rat following release from duct ligation. J Mol Histol 2005;36:199–205.
108 Takahashi S, Schoch E, Walker NI: Origin of acinar cell regeneration after atrophy of the rat parotid induced by duct obstruction. Int J Exp Pathol 1998;79:293–301.
109 Harrison JD, Garrett JR: Histological effects of ductal ligation of salivary glands of the cat. J Pathol 1976;118:245–254.
110 Osailan SM, Proctor GB, McGurk M, Paterson KL: Intraoral duct ligation without inclusion of the parasympathetic nerve supply induces rat submandibular gland atrophy. Int J Exp Pathol 2006;87:41–48.
111 Shiba R, Hamada T, Kawakats K: Histochemical and electron microscopical studies on effect of duct ligation of rat salivary glands. Arch Oral Biol 1972;17:299–309.
112 Cotroneo E, Proctor GB, Paterson KL, Carpenter GH: Early markers of regeneration following ductal ligation in rat submandibular gland. Cell Tissue Res 2008;332:227–235.
113 Emmelin N, Garrett JR, Ohlin P: Secretory activity and the myoepithelial cells salivary gland after duct ligation in cats. Arch Oral Biol 1974;19:275–283.
114 Takahashi S, Nakamura S, Suzuki R, Domon T, Yamamoto T, Wakita M: Changing myoepithelial cell distribution during regeneration of rat parotid glands. Int J Exp Pathol 1999;80:283–290.
115 Takahashi S, Shinzato K, Domon T, Yamamoto T, Wakita M: Mitotic proliferation of myoepithelial cells during regeneration of atrophied rat submandibular glands after duct ligation. J Oral Pathol Med 2004;33:430–434.

116 Burgess KL, Dardick I, Cummins MM, Burford-Mason AP, Basset R, Brown DH: Myoepithelial cells actively proliferate during atrophy of rat parotid gland. Oral Surg Oral Med Oral Pathol Oral Radiol Endod 1986;82:674–680.
117 Takahashi S, Nakamura S, Shinzato K, Domon T, Yamamoto T, Wakita M: Apoptosis and proliferation of myoepithelial cells in atrophic rat submandibular glands. J Histochem Cytochem 2001;49: 1557–1563.
118 Takahashi S, Shinzato K, Domon T, Yamamoto T, Wakita M: Proliferation and distribution of myoepithelial cells during atrophy of the rat sublingual gland. J Oral Pathol Med 2002;32:90–94.
119 Martinez JR, Bylund DB, Cassity N: Progressive secretory dysfunction in the rat submandibular gland after excretory duct ligation. Arch Oral Biol 1982;27:443–450.
120 Hoffman MP, Kidder BL, Steinberg ZL, Lakhani S, Ho S, Kleinman HK, Larsen M: Gene expression profiles of mouse submandibular gland development: FGFR1 regulates branching morphogenesis in vitro through BMP- and FGF-dependent mechanism. Development 2007;129:5767–5778.
121 Chiles TC: Regulation and function of cyclin D2 in B lymphocyte subsets. J Immunol 2004;173:2901–2907.
122 Supp DM, Witte DP, Branford WW, Smith EP, Potter SS: Sp4, a member of the Sp1 family of zinc finger transcription factors, is required for normal murine growth, viability, and male fertility. Dev Biol 1996; 176:284–299.
123 Leimeister C, Bach A, Gessler M: Developmental expression patterns of mouse sFRP genes encoding members of the secreted frizzled-related protein family. Mech Dev 1998;75:29–42.
124 Xu QH, D'Amore PA, Sokol SY: Functional and biochemical interactions of Wnts with FrzA, a secreted Wnt antagonist. Dev Biol 1998;198:219.
125 Sadler KC, Krahn KN, Gaur NA, Ukomadu C: Liver growth in the embryo and during liver regeneration in zebrafish requires the cell cycle regulator, uhrf1. Proc Natl Acad Sci USA 2007;104:1570–1575.
126 Bronner C, Achour M, Arima Y, Chataigneau T, Saya H, Schini-Kerth VB: The UHRF family: oncogenes that are drugable targets for cancer therapy in the near future? Pharmacol Ther 2007;115:419–434.
127 Akkiprik M, Feng YM, Wang HM, Chen KX, Hu LM, Sahin A, Krishnamurthy S, Ozer A, Hao XS, Zhang W: Multifunctional roles of insulin-like growth factor binding protein-5 in breast cancer. Breast Cancer Res 2008;10.

Guy H. Carpenter
Salivary Research Unit, Floor 17, Tower Wing
King's College London Dental Institute, London SE1 9RT (UK)
Tel. +44 207 188 7460, Fax +44 207 188 7458
E-Mail guy.carpenter@kcl.ac.uk

Tucker AS, Miletich I (eds): Salivary Glands. Development, Adaptations and Disease.
Front Oral Biol. Basel, Karger, 2010, vol 14, pp 129–146

Salivary Gland Disease

Bethan L. Thomas[a] · Jackie E. Brown[a] · Mark McGurk[b]

[a]Department of Dental Radiology, and [b]Department of Oral and Maxillofacial Surgery, King's College London Dental Institute, London, UK

Abstract

Salivary gland disease covers a wide range of pathological entities, including salivary gland-specific disease, as well as manifestations of systemic diseases. This chapter discusses the recent advances in managing obstructive salivary gland disease, the move from gland excision to gland preservation, the dilemmas in diagnosing and managing tumours of the salivary glands, and the international data collection to understand the aetiology and progression of Sjögren's disease.

Salivary glands in humans include the paired parotid glands, paired submandibular glands, paired sublingual glands (located in the floor of the mouth) and hundreds of minor salivary glands distributed around the mouth (lips, cheeks, tongue, floor of mouth, palate) all of which may be affected by disease. In some diseases all glands may be affected while in other diseases only a single gland is involved. The latter tend to be salivary-specific disease (e.g. salivary tumours, obstructive disorders, infections), and the former systemic disorders (e.g. Sjögren's disease, lymphoma, sarcoidosis and metabolic diseases).

In salivary disease the history of the presentation on its own is a good pointer to diagnosis, and to a lesser extent clinical examination (in obstructive disease there may be little clinical evidence of the disease), and therefore in most cases the diagnoses is decided by special investigations. Since most of these are not readily available to general practitioners the diagnosis and management of salivary gland disease lies principally in the domain of specialists within the hospital environment. Investigations include ultrasound, sialography, magnetic resonance imaging (MRI) and endoscopy.

The spectrum of disease encountered in salivary glands mimics that seen in other glands: neoplasms, inflammatory disorders (including infection and obstruction), autoimmune disease (principally Sjögren's disease) and systemic disease manifesting in the gland (e.g. HIV). This chapter will not attempt to cover the full panoply of

Table 1. Overview of salivary gland disease

Group	Subgroup		Ref.
Inflammatory disease	Obstructive disease	Classically presents as meal-time syndrome Caused by calculi (stones), strictures or mucus plugs	6
	Recurrent parotitis of childhood	Develops in childhood Chronic recurrent swelling of parotids, unilateral or bilateral May persist into adulthood Aetiology is poorly understood	40, 41
	Sialadenitis	Bacterial Acute and chronic infections Viral, e.g. paramyxovirus (mumps)	42, 43
	Necrotizing metaplasia	Poorly understood aetiology Most frequently affect minor salivary gland on the palate Masquerades as a tumour	
Tumours	Benign	See table 2	44
	Malignant	See table 2	44
Autoimmune	Sjögren's	See section in this chapter	38, 45
	SOX	Sialadenitis, nodal osteoarthritis and xerostomia	46
Mucocoele	Minor gland	Small submucosal swelling at site of minor salivary gland, classically bluish colour Self-limiting or surgical removal	
	Sublingual gland	Ranula in floor of mouth (mucocele arising from sublingual gland): Swelling in floor of mouth Varies in size, up to several centimetres a. Remove affected gland or b. Marsupialize	47, 48
	Sublingual gland	Plunging ranula May have very few intraoral symptoms, just a fluctuant swelling in the neck Mucocele passes through mylohyoid resulting in swelling in submandibular or submental triangle Removal of affected gland	49
Xerostomia	Drug-induced	e.g. tricyclic antidepressants	50
	Age-related		51
Sialosis		Enlargement of salivary gland(s), but with normal tissue on radiographic, ultrasound and histological analysis Associated with endocrine disorders such as diabetes, liver disease, and also with bulimia	52, 53
Developmental abnormalities	Aplasia Atresia		54–56

conditions (table 1) but rather the three most common disorders, notably obstructive disease, neoplasms, and Sjögren's disease.

Salivary Gland Obstructive Disease

The most significant advances in salivary gland disease management in the last two decades have occurred with obstructive disease. This typically presents with a history of intermittent obstruction, interspersed with long periods (months or years) of normal function. The symptoms classically occur at meal times, when salivary secretion is increased and unable to pass the obstruction with resultant build up of saliva resulting in inflation of the gland and ductal system. The swelling usually subsides within a matter of hours, as the saliva seeps past the stone or stricture, leaving no trace of the previous swelling, and therefore often baffling patients. Such symptoms may occur sporadically; the mean time from onset of mealtime syndrome to presentation in hospital is about 5 years. The latter is usually precipitated by an episode of subacute sialadenitis (inflammation of the salivary gland). Paradoxically, patients with large stones (≥1 cm diameter) may have no history of obstruction if the stone has pushed out of the main part of the duct to lie in a diverticulum. These stones present with an episode of sialadenitis.

Obstructive salivary gland disease is thought to account for approximately half of major salivary gland disorders [1]. The incidence of salivary gland stones in the UK population has been estimated from NHS admission data to be 60 patients per million population per annum [2]. In deriving the figures for prevalence, assumptions have been made that most of the admissions for sialadenitis are related to salivary gland calculi. However, this is an underestimate of the problem because a retrospective review of sialograms performed over a 10-year period at Guy's and St Thomas' NHS Trust revealed that 23% of obstructed cases were attributable to strictures, and 4% to mucus plugs, the rest being attributable to stones [3]. These conditions do not tend to lead to sialadenitis and admission to hospital, so the true estimate of the problem is closer to 75 cases per million population per annum.

Until recently, patients with sialadenitis had little choice regarding treatment. A stone close to the orifice of the duct could be released easily under local anaesthesia but other stones were treated by gland removal. These procedures risked injury to the branches of the facial nerve with its associated sequelae (fig. 1), plus gustatory sweating (Frey's syndrome).

Salivary Gland Calculi

There is now clear evidence that 80% of stones can be retrieved leaving an asymptomatic functional salivary gland in situ. Both animal [4] and human [5] studies of salivary excretion have demonstrated that gland function improves following release of obstruction. This is corroborated by the appearance of sialograms following stone removal where improved clearance of contrast medium is observed compared to

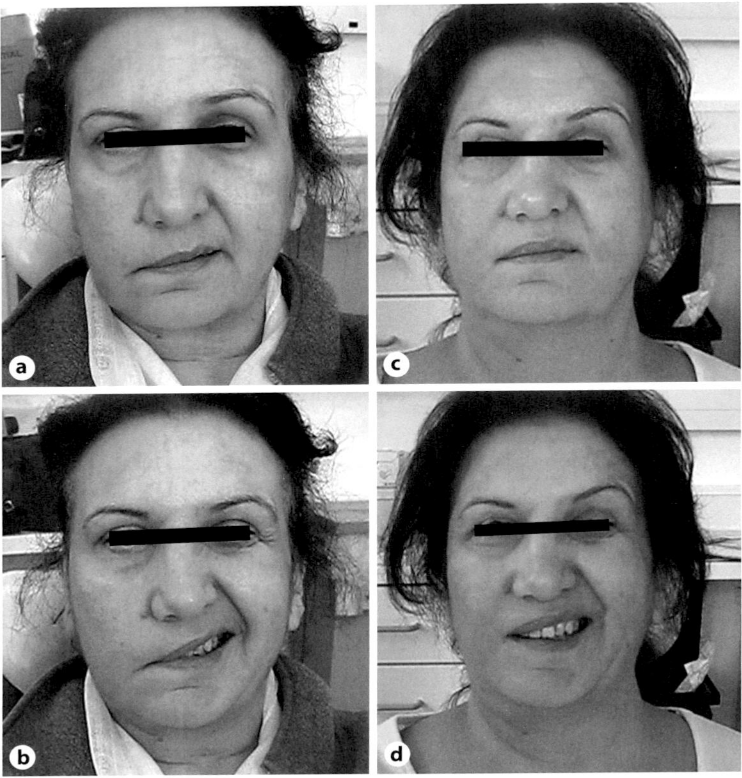

Fig. 1. Right facial nerve palsy following surgical removal of a tumour from the right parotid gland. At 1 month post-surgery the aesthetic effects of loss of tone within the muscles of facial expression at rest (**a**) and in function (**b**) are clearly seen. Limited recovery of function was seen at 3 months post-surgery following the use of TENS (**c**, **d**).

Fig. 2. Plain radiograph film showing the presence of 7 stones along the left submandibular gland duct within the floor of mouth. The anterior 2 stones would be amenable to simple surgical release, without significant morbidity, but the remaining stones would require more extensive surgery to open the floor of mouth, and avoid the lingual nerve [6, 19]. Alternatively, use of basket retrieval via interventional sialography or endoscopic guidance is likely to be effective given the size of the stones.

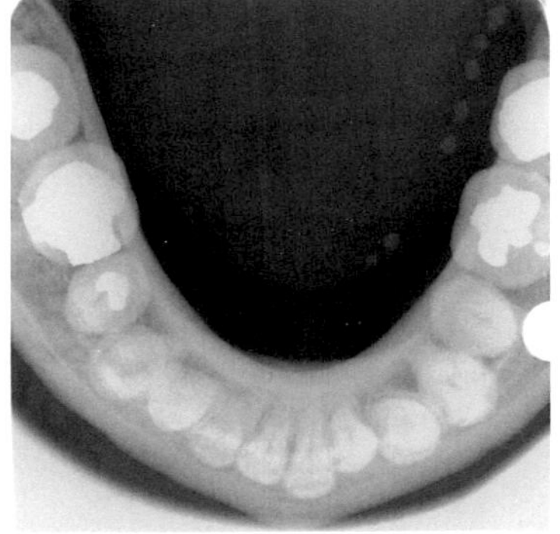

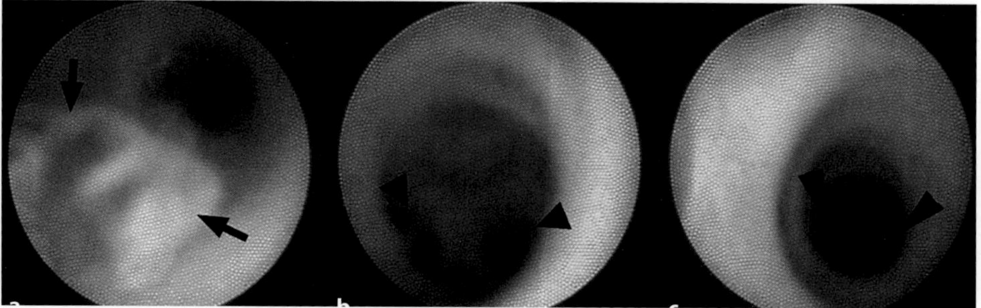

Fig. 3. Images from an endoscopic examination showing debris within the duct (**a** arrows), and the capability of the current endoscopes to reach the secondary (**b** arrowheads) and tertiary ducts (**c** arrowheads) of the salivary glands.

Table 2. Current management protocol for stone removal by minimally invasive techniques

Stone location	Stone size and situation	First-line treatment
Any gland. Mobile stone within a duct	≤4 mm	Basket retrieval (radiological guidance or endoscopy)
Submandibular	>4 mm Large and fixed	Surgical removal
Parotid	>4 + <8 mm and fixed	Extracorporeal lithotripsy
Parotid	>8 mm	Endoscope guided surgery

pre-treatment sialographs, suggesting improved gland function. There are therefore clearly demonstrable capabilities of the salivary glands to recover and regenerate, which has created the stimulus to move away from sialadenectomy and move to gland-preserving treatment.

Treatment Modalities for Stone Removal

The three treatment options for proximal salivary calculi, not amenable to simple surgical release, are basket retrieval (via endoscopic, fluoroscopic sialography, or ultrasound guidance) [6–10], lithotripsy [11–17] or surgical removal (with endoscopy guidance for parotid stones) [18–22]. The decision as to which modality to use is dependent on several parameters. Size and position of a stone are important factors. A calculus in the gland parenchyma or small secondary ducts cannot be engaged by a basket. Stones in the submandibular duct that lie over the posterior border of mylohyoid, beyond the genu (the bend in the duct as it nears the hilum) are difficult to reach. Size is a factor as it must be possible to retrieve the stone down the duct. Small, mobile stones are the targets for basket retrieval (fig. 2–4; table 2).

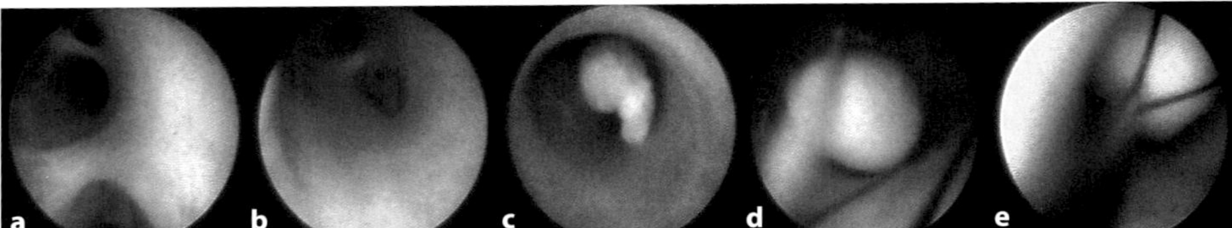

Fig. 4. Images from an endoscopic-guided basket retrieval of a stone. **a**, **b** Primary and secondary ducts distal to the position of the stone. **c** Stones are visualised. **d, e** Capture of the stones within a basket which is then closed around the stone (**e**) to allow withdrawal of the stone from the duct in the basket.

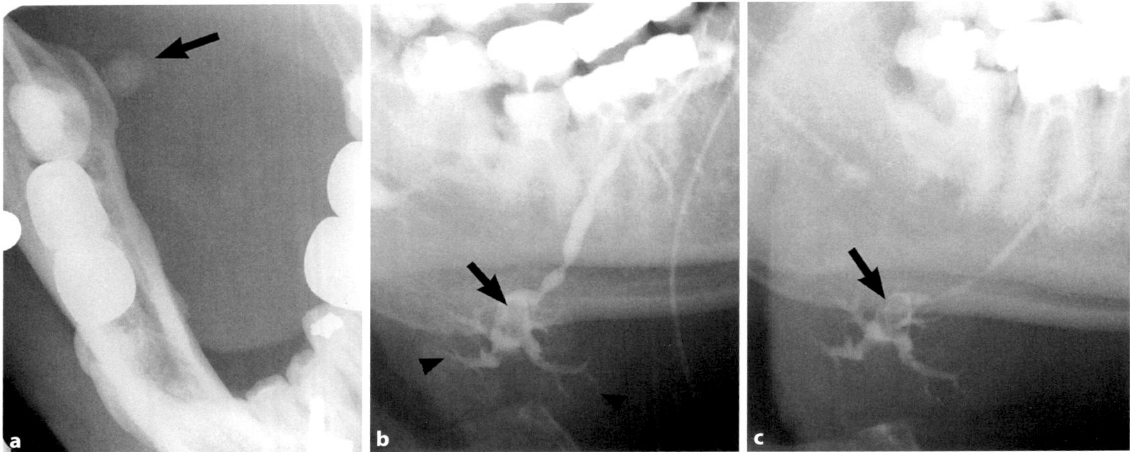

Fig. 5. a Plain radiographic lower posterior oblique occlusal view showing the presence of a large stone at the posterior of the right floor of mouth. **b** Plain film sialogram using iodine-based contrast medium, revealing that this large stone is present in the hilum of the right submandibular gland, as exhibited by the filling defect marked by the arrow; numerous strictures lie distal to the stone, within the main duct. **c** A further image taken 1 min after removal of the cannula from the duct orifice to allow the gland to empty. The retention of contrast in the ductal system is not normal and indicates the obstructive nature of this stone preventing full clearance of contrast, and therefore also saliva, from the gland. The size and position of this stone dictate that it is not amenable to either lithotripsy or basket retrieval, but instead requires surgical removal of the stone via an intraoral approach.

The treatment of large and/or fixed stones depends on which gland is involved. Parotid stones are amenable to extracorporeal shock-wave lithotripsy. High-energy ultrasound waves are used to fragment the stone, in a similar manner to kidney stones. Lithotripsy is not very effective for submandibular gland stones. This is partly due to the difficulty of targeting the stone, but also because submandibular stones can be larger than those in the parotid. Size of the stone has an important bearing on success of lithotripsy. A stone >8 mm has less than a 10% chance of clearance (table 2).

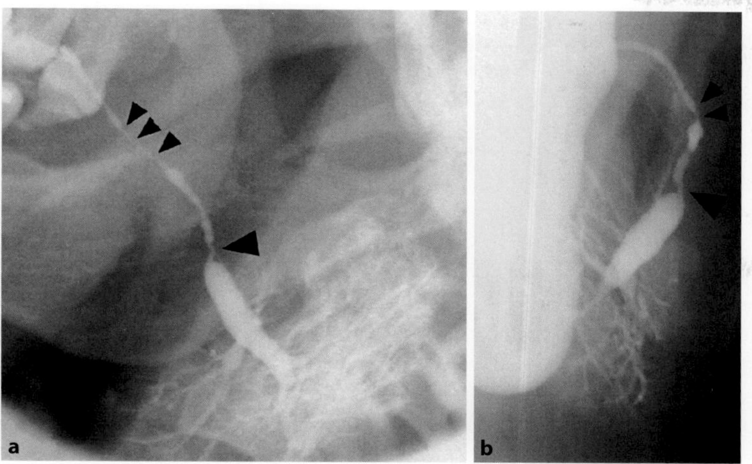

Fig. 6. Sialogram showing the presence of a long diffuse stricture of the left parotid main duct (**a**) shown on a lateral oblique view and (**b**) on an anterior-posterior view with dilatation of the duct proximal to the stricture.

The final option for submandibular stones >4 mm in diameter and for parotid stones that fail with lithotripsy is gland-preserving surgery [7, 17–19, 21]. Submandibular gland stones have a propensity to develop at the genu of the duct (fig. 5), near to the lingual nerve. The techniques developed to retrieve the stone require a long incision from the duct orifice at the anterior of the floor of the mouth to the wisdom tooth. By raising the sublingual gland and rotating it laterally, the duct can be traced to the hilum of the gland and the stone. This procedure has low morbidity and preserves the gland.

The surgical approach to parotid stones is via an extraoral skin incision in the preauricular region. An endoscope is placed within the duct and directs the surgeon onto the stone. Minimal dissection of the parotid gland reduces the risk to the facial nerve.

Management of Strictures

Strictures occur in different forms (single-point strictures, multiple strictures, diffuse strictures) (fig. 6) and their aetiology is not understood. Treatment methods have been adapted from the balloon angioplasty techniques used to dilate heart vessels in vascular interventional procedures. A collapsed balloon is directed into the stricture over a guide wire. The procedure can be performed under sialography or endoscopic guidance. The balloon is then filled with saline under pressure to stretch the duct and some operators routinely use stents to maintain duct patency following dilatation [23]. From the early data there seems to be at least 92% success rate [24]. Some lesions reform over time and multiple procedures may be necessary to avoid gland excision.

Table 3. Frequency of tumours[1]

	Frequency of tumours per site	Risk of malignancy at each site	Nottingham series; frequency of tumours per site	Nottingham series; risk of malignancy at each site
Parotid	100	<25%	100	6%
Submandibular	10	50%	11	34%
Sublingual	1	100%	1	100%
Minor gland	10	>80%	10	24%

[1] This table indicates the generally accepted frequency of tumours between the salivary glands and risk of malignancy for a given tumour at that site [31]. This is compared to a series collected from Nottingham, UK, based on data derived from a histopathology database. This revealed the frequency of tumours at each site to be quite accurate to the predicted frequency, but with a reduced risk of malignancy in the minor salivary glands than predicted [27].

A recently published report combining the minimally invasive results of 5 centres (1990–2004) (Milan, Paris, Erlanger, London, Israel) reports an overall success rate of 80% [25], and a gland removal rate of only 3%. This success was achieved using a range of the minimally invasive techniques described above. This represents a significant change in management from the current standard of practice which is almost 100% gland excision to over 97% gland preservation.

Tumours of the Salivary Glands

Incidence

Salivary gland tumours are an uncommon entity compared to other tumours in the body (0.3% of all malignant tumours in the USA) [26]. An accurate incidence for both benign and malignant disease is difficult to establish. Institutional data contains bias due to the nature of the specialist interests of such institutions. Comparison of incidence across different populations globally has also proved problematic due to the lack of uniformity by which information is collected in different countries. This dilemma has been solved by a population-based study undertaken in Nottingham, UK [27]. It showed that the incidence of benign salivary tumours (in all glands) is 80 cases per million population annually and in the same population 7 cancers will develop. Despite the difficulties in collecting accurate, comparable data there does appear to be a slight increase in prevalence of salivary gland tumours within certain populations [28–30]. What is clear is that in all populations the incidence of these cancers increases with age, and with increasing age there is an increased incidence in men. The average age of presentation of benign tumours precedes that of malignant tumours by a decade (46 and 57 years).

Distribution by Site of Benign and Malignant Salivary Gland Tumours

It has been recognised for over a century that there are clear differences in the frequency of presentation of various tumours within each salivary gland. There is a significantly higher incidence of benign tumours compared to malignant. The data shows that for every 100 parotid tumours, there are 10 SMG tumours, 10 tumours involving a minor salivary gland and 1 involving a sublingual gland [31]. The relative risk of encountering malignancy is 10–15% in the parotid, 30–40% in the submandibular gland and soft palate minor glands, but is nearly 100% for tumours of the minor glands in the floor of mouth or involving the sublingual gland (table 3).

In clinical terms, these relative frequencies of occurrence between the salivary glands means that a clinician is most likely to encounter tumours in the parotid gland, the majority of which will be benign, most often pleomorphic adenomas (60%), followed by Warthin's tumours (20%). About 40–50% of salivary cancers are indolent in nature and in the early stages of development can masquerade as benign lumps. This is particularly so with minor salivary gland cancers which history proves are readily mistaken for the pathology of other conditions (cysts, fibromas, scar tissue). The major salivary glands have a different embryologic origin and their cellular constituents (serous and mucous) differ; consequently it is understandable that different glands will be susceptible to different tumour types.

Histological Tumour Types

The histological analysis of salivary gland tumours is complex, with the most recent WHO classification distinguishing 42 tumours of salivary glands (excluding metastatic deposits), of which there are 24 malignant epithelial tumours and 13 benign epithelial tumours (table 4) [26]. This huge array of histological varieties may be a feature of the varying cellular contributions of these differing major and minor salivary glands. Animal experiments demonstrate that salivary tissue has the potential for significant diversity (hence the name pleomorphic adenoma). The reason for this characteristic is unclear. Unfortunately the new histological classification does not help to distinguish the behaviour that a particular tumour is likely to exhibit. Given that salivary gland malignancies represent less than 1% of all malignancies, the 24 epithelial malignant variations of salivary gland tumours described are therefore very infrequent; assessment of the behaviour of these tumours is going to take considerable time to establish. All molecular information and any understanding of the cellular origin of the tumour and cellular behaviour within the tumour will be beneficial in defining more precisely the histological diagnosis of the various tumours. Such information will help in the prediction of behaviour, and thus may be beneficial in the future, guiding clinical decisions regarding treatment.

Management

As stated above, the most frequently encountered tumour is a pleomorphic adenoma. These present as solitary, firm, rubbery lumps within the parotid gland, which is fairly characteristic so most experienced salivary gland surgeons would predict the

Table 4. WHO classification of salivary gland tumours (adapted from Eveson et al. [26])

Malignant epithelial tumours	Acinic cell carcinoma Mucoepidermoid carcinoma Adenoid cystic carcinoma Polymorphous low-grade adenocarcinoma Epithelial-myoepithelial carcinoma Clear cell carcinoma, not otherwise specified Basal cell adenocarcinoma Sebaceous carcinoma Sebaceous lymphadenocarcinoma Cystadenocarcinoma Low-grade cribriform cyst adenocarcinoma Mucinoous adenocarcinoma Oncocytic carcinoma Salivary duct carcinoma Adeoncarcinoma, not otherwise specified Myoepithelial carcinoma Carcinoma ex pleomorphic adenoma Carcinosarcoma Metastasizing pleomorphic adenoma Squamous cell carcinoma Small cell carcinoma Large cell carcinoma Lymphoepithelial carcinoma Sialoblastoma
Benign epithelial tumours	Pleomorphic adenoma Myoepithelioma Basal cell adenoma Warthin tumour Oncocytoma Canalicular adenoma Sebaceous adenoma Lymphadenoma sebaceous Lymphadenoma non-sebaceous Ductal papilloma – inverted ductal papilloma Ductal papilloma – intraductal papilloma Ductal papilloma – sialadenoma papilliferum Cystadenoma
Soft tissue tumours	haemangioma
Haematolymphoid tumours	Hodgkin's lymphoma Diffuse large B-cell lymphoma Extranodal marginal zone B-cell lymphoma
Secondary tumours	

diagnosis accurately. For all suspected tumours, prior to definitive treatment, investigations are performed, including imaging, and sampling for histological analysis. Imaging can typically include ultrasound, MRI and CT (computerised tomography), which help to define the extent of a lesion and in particular, whether it extends into the deep lobe of the parotid.

Histological assessment is crucial prior to treatment of any tumour; this aims to distinguish benign from malignant disease, and further determine a histological diagnosis allowing prediction of tumour behaviour, and therefore inform on the extent of treatment required. The parotid and submandibular glands pose specific problems due to their relationship with the facial nerve, most significantly so for the parotid gland. Since the incidence of facial nerve injury increases significantly with a second surgery to a gland it is not possible to take an open biopsy for histopathological diagnosis ahead of definitive surgery. Damage to the facial nerve presents a significant aesthetic and functional problem (fig. 1). This cannot always be avoided, particularly in cases where a tumour is in close proximity to a branch of the facial nerve, or even the main trunk, but knowing the increased risk of nerve damage with repeat surgery, treatment must aim for adequate management by eradication of the tumour during primary surgery. In view of the need to avoid repeat surgery, diagnosis relies heavily on fine-needle aspiration biopsy (FNA), known to have only 60–75% full diagnostic accuracy [32], and its inherent false-positive and false-negative results. This creates a problem whereby a surgeon may enter the surgical procedure without an absolute diagnosis, and may, based on the combination of FNA and imaging results, just have an indication as to whether a lesion is benign or malignant, and its position. In any other surgical field this might prove less problematic as the balance of risk of morbidity to the benefit of eradication of tumour and thus cure may allow more extensive surgery to be feasible, balanced against loss of function/aesthetics. In the field of the parotid gland however, given the relationship with the facial nerve this is not possible.

The standard procedure for removal of benign tumours presenting in the superficial part of the parotid gland was traditionally superficial parotidectomy; this procedure resulted in an aesthetic deformity. To circumvent such deformity, surgeons have developed techniques to reduce the extent of surgery for benign tumours, thus preserving gland tissue and maintaining aesthetics. The procedure currently adopted is called extracapsular dissection, whereby surgery is performed outside the capsule, defining a layer within the normal gland around the tumour. This method maintains the integrity of the capsule and avoids spillage of the tumour. A potential hazard with such an approach is an increased risk of tumour recurrence, either from satellite lesions from the capsule being left allowing a further tumour to form, or, if misdiagnosis of a malignant tumour had occurred preoperatively, it may have been inadequately excised. Use of frozen sections at the time of surgery can complement the preoperative FNA analysis. While this is a good predictor of benign disease it is less accurate at diagnosing malignant disease (85.9% accurate) [33]. This technique therefore risks underdiagnosis of malignancy, with the risk of undertreating and the consequent sequelae.

However, in spite of these potential risks there is clear evidence of the efficacy of the use of the technique of extracapsular dissection without deleterious effects based on the findings of a study looking at 5- and 10-year cancer-specific survival rates for superficial parotidectomy compared to extracapsular dissection. In this well-documented surgical series, from the Christie Hospital in Manchester, surgeons made decisions to proceed to total parotidectomy, superficial parotidectomy or extracapsular dissection, based on clinical analysis of the lump, rather than the use of preoperative FNA. For cases where the lump was <4 cm, showed mobility, with no involvement of the facial nerve and no involvement of the deep lobe, the surgeons used a more conservative management approach [34]. 2.4% of tumours treated by extracapsular dissection were found to be malignant tumours on histological analysis. Analysis of the 5- and 10-year cancer-specific survival rates for this group of patients (malignant tumour removed by extracapsular dissection) was found to be 100%, comparable to that of the group with malignant tumours treated by superficial parotidectomy (98%) [35]. This data thus indicates the safety of the use of this technique for experienced surgeons.

The data also significantly revealed the role of clinical experience in diagnosis of these tumours. The surgeons involved in this series predicted the nature of a lesion (i.e. benign vs. malignant lesions) based on clinical assessment of size, texture, and fixedness of a lump correctly in 97.6% of cases, which compares favourably to reported prediction values for FNA of 81–98% [26]. This suggests that there is a small proportion of salivary gland tumours that are difficult to diagnose not only at the cellular, histological level, but also masquerade clinically as benign lesions. This may elude to the propensity for some malignant lesions to behave more like benign lesions, and hence the small number of malignant tumours treated by extracapsular dissection in this series. While we currently have the histological abilities to describe morphological specificities of a wide variety of salivary gland tumours, our understanding of the behaviour of those tumours at the molecular, cellular and tissue level is lacking. A better understanding of the molecular and cellular variation amongst salivary tumours will allow a better insight into the cellular behaviour and capabilities of the various types of tumour. This knowledge would allow molecular biology testing of the small number of cells produced by FNA, producing a more accurate assessment of the tumour. This would go a considerable way to providing accurate presurgical diagnosis and further reduce the need for postoperative radiation to manage those tumours incorrectly diagnosed preoperatively as benign.

Sjögren's Syndrome

Sjögren's syndrome (SS) is an autoimmune disease of connective tissue. It affects not only salivary glands but other glands as well, including the lacrimal glands resulting in dry eyes and damage to the cornea. In addition there are other systemic effects, including pulmonary and vascular problems and fatigue and joint pain. While the effects on the salivary glands involve reduction in function and hence reduced salivary flow, with the concomitant risk of increased caries rate and other infections, the

Table 5. EU-USA classification criteria for Sjögren's syndrome [37]

I	Symptomatic xerostomia for >3 months, persistently swollen salivary glands as an adult, or frequent use of liquids to aid in swallowing food
II	Symptomatic dry eyes for >3 months, recurrent 'grittiness', the use of tear substitutes >3 times per day
III	Positive Schirmer's test or rose Bengal score (or other ocular dye score, e.g. lissamine green)
IV	Abnormal lower lip biopsy (focus score ≥1)
V	Positive result for unstimulated whole salivary flow (≤1.5 ml in 15 min), or other objective test of changes in salivary glands (e.g. a sialogram showing sialectasis in the absence of obstruction)
VI	Antibodies to Rho (SSA) or La (SSB), or both
	Exclusions: any patient with past head and neck radiation treatment, hepatitis C infection, acquired immunodeficiency disease (AIDS), pre-existing lymphoma, sarcoidosis, graft vs. host disease, use of anticholinergic drugs
1°	Four of the six criteria must be positive, but this must include either histological evidence of minor salivary gland involvement (IV) or serological evidence (VI) For asymptomatic patients, three of the four objective criteria must be met (III, IV, V, VI)
2°	1. The presence of other connective tissue (CT) disease 2. Either symptomatic dry mouth (I) or symptomatic dry eyes (II), *plus* any two of the objective tests should be present *excluding* serology (as this may be present as part of the other CT disease) Patients should therefore exhibit objectively assessed dry eyes (III), or histological evidence of Sjögren's (IV), or objective dry mouth (V)

systemic manifestations are also of critical significance to the health of the patient. Further, the massive infiltration of lymphocytes results in a predisposition to development of MALT lymphoma. These health issues combined are sufficient to warrant extensive research into this disease.

SS affects 3–4% of adults in the UK, of which 90% are female, and the mean age of diagnosis is 50 years [36]. The aetiology is currently not well understood and the range of symptoms (and the severity) any individual experiences can vary. There is no single test for SS, rather a classification based on a series of subjective and objective criteria [37]. The current diagnostic criteria are shown in table 5, but it should be noted that the diagnostic criteria for SS have been changed numerous times in the last four decades, and there is still debate and disagreement regarding the current classification.

SS is subdivided into primary and secondary SS. Primary Sjögren's presents with no other associated connective tissue disease, whereas secondary Sjögren's is defined by a patient having another connective tissue disease, such as rheumatoid arthritis or systemic lupus erythematosus. The difficulty in defining the diagnostic parameters is that not all patients exhibit the same manifestations of the disease, for example not all patients carry either of the SS antibodies SSA (anti-Rho) and SSB (anti-La), while others are positive for one or the other. It has been shown previously that 70% of Sjögren's patients are positive for the SSA antibody, and 40% are positive for the SSB antibody. Since these antibodies can also be found in patients with systemic lupus erythematosus, it adds to the confusion in diagnosis and leads to a poor understanding of the nature of the disease and its distinction from other connective tissue autoimmune disorders. Anti-SSA and anti-SSB are known to be specific to RNA-protein complexes, but it is not clear how this leads to the disease, and since these complexes are present in all cells does not explain why there are specific target tissues. It is clear that many people have high levels of these autoantibodies, and other autoimmune antibodies such as antinuclear antibody, yet do no proceed to develop significant disease, if any.

One of the major issues in trying to study this disease, with its subclasses and controversial diagnosis, is the ability to select a suitably large cohort to make any significant assessment. Due to this problem, an international group, called SICCA (Sjögren's International Collaborative Clinical Alliance) is currently recruiting patients to a global project to run a longitudinal study. Patients are selected from a large remit with symptomatic dry mouth or dry eyes and also those with proven Sjögren's, to formally diagnose their disease based on the current criteria and to follow the changes occurring in terms of symptoms, serology and histologically within the minor salivary glands over a 2-year period [38]. The long-term aim is to classify Sjögren's disease and its subsets more precisely, and determine whether the current diagnostic criteria are suitable and adequate. The first data from this project was recently published [39].

Over 1,200 patients had been recruited to the study by the time the first data was analysed. An apparently simple comparison of patients' subjective assessment of dry mouth or eyes, based on a simple question of whether they felt dry, did not correlate to levels of serum anti-SSA and anti-SSB, lymphocytic infiltration counts in the salivary glands or ocular staining measurements. A subjective assessment, albeit with functional significance, did produce some association to the objective measures. For example, by asking if a drink is required to be able to swallow dry foods (which may be a better gauge of the functional production of saliva) there was some correlation to the objective measurements. Given the size of the cohort that is planned it will hopefully become apparent how closely these objective measures associate with patients' functional saliva production.

The significant finding to date is in the assessment of the proportion of patients presenting with the various combinations of three of the objective criteria of SS, namely ocular staining, histological assessment of labial gland, and presence of anti-SSA and anti-SSB antibodies. A distinct group of patients was identified, presenting with significant objective dryness of eyes, but neither of the other two criteria (serum levels of anti-SSA and anti-SSB, and lymphocytic infiltration of minor salivary gland). They made up a similar size group to those presenting with positive results for all three criteria. Since this group, with only keratoconjunctivitis sicca, demonstrate lower levels of serum autoantibodies they will be an interesting group to follow to determine their disease progression and correlation with the presence of other serum factors, such as antinuclear antibody, and later production of anti-SSA and anti-SSB.

At the 2-year follow-up stage, only 2% of the patients revealed significant progression of their disease. This highlights the difficulty in obtaining a significantly sized cohort to adequately assess the progression of the disease. It will still be many years before the full results from this study become available, but it is a necessary worthwhile wait to provide the data needed.

Discussion

This chapter has attempted to give some insight into the dilemmas of diagnosis and treatment of salivary gland tumours, as well as the most common autoimmune disease affecting salivary glands, Sjögren's syndrome. Both areas are in need of further basic research to understand the disease processes and lead to better management. The chapter has also discussed the significant improvement in management of salivary gland obstructive disease that has occurred over the last two decades. This work has clearly identified the ability of glands to recover and regenerate, but the limits of this regeneration are not understood. One group that could benefit from further research into this area are patients with dry mouth attributable to age-related degenerative changes. It has been established that saliva production decreases with age. With extended life expectancies and the associated expansion of the population in the eighth, ninth and tenth decade, this will become a more significant clinical problem for healthcare providers. Within this population (who have increasingly high expectations for quality of life) there will be a need to manage the associated clinical problems relating to dry mouth (xerostomia), namely increased rates of caries, periodontal disease and other oral infections such as candidiasis, and unpleasant symptoms including discomfort, loss of taste, difficulty with speech and eating. It is therefore imperative to determine better methods to manage xerostomia than the current limitations of symptomatic relief with topical medication and sialogogues. The recent knowledge gained in our understanding of gland recovery following release of obstruction will hopefully lead to a better understanding of how such recovery occurs at the cellular and molecular level, and allow inroads into the development of methods to stimulate salivary gland regeneration in the remaining functional salivary tissue of patients with xerostomia. An understanding of the pool of stem cells in this elderly population is vital to allow developments along this route.

One other area of interest that has been overlooked is sialosis. In sialosis excessive salivary gland tissue is produced which appears normal by imaging and histological analysis. Patients presenting with this are primarily concerned with the significant effect it has on their facial appearance as a result of the extensive gland enlargement, usually only affecting the parotid glands. In such cases the desire to reduce the size of the salivary glands is primarily driven by aesthetic reasons, since the glands do not exhibit any pathology requiring intervention. Little is known of the aetiology of this disease, and there are currently no treatments to manage it effectively. This phenomenon may however present an ideal opportunity, currently being overlooked, to harness the molecular pathways driving this gland generation and use it to stimulate salivary gland regeneration in patients with xerostomia. This alone will not provide answers to address salivary gland regeneration, but coupled with information gained from embryological studies and obstruction and regeneration studies, will hopefully help to identify improved treatment for xerostomia.

References

1 Epker BN: Obstructive and inflammatory diseases of the major salivary glands. Oral Surg Oral Med Oral Pathol 1972;33:2–27.

2 Escudier MP, McGurk M: Symptomatic sialoadenitis and sialolithiasis in the English population, an estimate of the cost of hospital treatment. Br Dent J 1999;186:463–466.

3 Ngu RK, Brown JE, Whaites EJ, Drage NA, Ng SY, Makdissi J: Salivary duct strictures: nature and incidence in benign salivary obstruction. Dentomaxillofac Radiol 2007;36:63–77.

4 Osailan SM, Proctor GB, Carpenter GH, Paterson KL, McGurk M: Recovery of rat submandibular salivary gland function following removal of obstruction: a sialometrical and sialochemical study. Int J Exp Pathol 2006;87:411–423.

5 Makdissi J, Escudier MP, Brown JE, Osailan S, Drage N, McGurk M: Glandular function after intraoral removal of salivary calculi from the hilum of the submandibular gland. Br J Oral Maxillofac Surg 2004;42:538–541.

6 McGurk M, Escudier MP, Thomas BL, Brown JE: A revolution in the management of obstructive salivary gland disease. Dent Update 2006;33:28–30.

7 McGurk M, Escudier MP, Brown JE: Modern management of salivary calculi. Br J Surg 2005;92:107–112.

8 Nahlieli O, Neder A, Baruchin AM: Salivary gland endoscopy: a new technique for diagnosis and treatment of sialolithiasis. J Oral Maxillofac Surg 1994;52:1240–1242.

9 Nahlieli O, Shacham, Bar T, Eliav E: Endoscopic mechanical retrieval of sialoliths. Oral Surg Oral Med Oral Pathol Oral Radiol Endod 2003;95:396–402.

10 Rogers DJ, Drage NA: Ultrasound-guided basket retrieval of a submandibular duct stone in a child. J Pediatr Surg 2006;41:1780–1782.

11 Iro H, Waitz G, Nitsche N, Benninger J, Schneider T, Ell C: Extracorporeal piezoelectric shock-wave lithotripsy of salivary gland stones. Laryngoscope 1992:102:492–494.

12 Iro H, Nitsche N, Waitz G, Schneider T, Benninger J, Ell C: Extracorporeal piezoelectric shock-wave lithotripsy of salivary gland stones: first clinical experiences. J Stone Dis 1992;4:8–12.

13 Zenk J, Bozzato A, Winter M, Gottwald F, Iro H: Extracorporeal shock-wave lithotripsy of submandibular stones: evaluation after 10 years. Ann Otol Rhinol Laryngol 2004;113:378–383.

14 Ottaviani F, Capaccio P, Campi M, Ottaviani A: Extracorporeal electromagnetic shock-wave lithotripsy for salivary gland stones. Laryngoscope 1996;106:761–764.

15 Ottaviani F, Capaccio P, Campi M: Parotid gland sialolithiasis: a new therapeutic option. Acta Otorhinolaryngol Ital 1997;17:58–63.

16 Capaccio P, Ottaviani F, Manzo R, Schindler A, Cesana B: Extracorporeal lithotripsy for salivary calculi: a long-term clinical experience. Laryngoscope 2004:114;1069–1073.

17 Katz P: New techniques for the treatment of salivary lithiasis: sialoendoscopy and extracorporeal lithotripsy: 1,773 cases. Ann Otolaryngol Chir Cervicofac 2004;121:123–132.

18 Capaccio P, Bottero A, Pompilio M, Ottaviani F: Conservative transoral removal of hilar submandibular salivary calculi. Laryngoscope 2005;115:750–752.

19 Combes J, Karavidas K, McGurk M: Intraoral removal of proximal submandibular stones – an alternative to sialadenectomy? Int J Oral Maxillofac Surg 2009;38:813–816.
20 McGurk M, MacBean AD, Fan KF, Sproat C, Darwish C: Endoscopically assisted operative retrieval of parotid stones. Br J Oral Maxillofac Surg 2006;44:157–160.
21 Thomas BL, Brown J, McGurk M: Advances in the treatment of salivary gland disease. Practitioner 2006;250:52–57.
22 Zenk J, Constantinidis J, Al-Kadah B, Iro H: Transoral removal of submandibular stones. Arch Otolaryngol Head Neck Surg 2001;127:432–436.
23 Nahlieli O, Shacham R, Yoffe B, Eliav E: Diagnosis and treatment of strictures and kinks in salivary gland ducts. J Oral Maxillofac Surg 2001;59:484–490.
24 Brown JE: Interventional sialography and minimally invasive techniques in benign salivary gland obstruction. Semin Ultrasound CT MR 2006;27: 465–475.
25 Iro H, Zenk J, Escudier MP, Nahlieli O, Capaccio P, Katz P, Brown J, McGurk M: Outcome of minimally invasive management of salivary calculi in 4,691 patients. Laryngoscope 2009;119:263–268.
26 Eveson JW, Auclair P, Gnepp DR, El-Naggar AK: Tumours of the salivary glands; in Barnes L, Eveson JW, Reichart P, Sidransky D (eds): World Health Organisation Classification of Tumours. Pathology and Genetics of Head and Neck Tumours. Lyon, IARC Press, 2005, pp 209–215.
27 Bradley P: General epidemiology and statistics in a defined UK population; in McGurk M, Renehan A (eds): Controversies in the Management of Salivary Gland Disease. Oxford, Oxford University Press, 2001, pp 3–12.
28 Schaefer O, Hildes JA, Medd LM, Cameron DG: The changing pattern of neoplastic disease in Canadian Eskimos. Can Med Assoc J 1975;112:1399–1404.
29 Hildes JA, Schaefer O: The changing picture of neoplastic disease in the Western and Central Canadian Artic (1950–1980). Can Med Assoc J 1984;130:25–32.
30 Friborg JT, Melbye M: Cancer patterns in Inuit populations. Lancet Oncol 2008;9:892–900.
31 Spiro RH: Management of malignant tumours of the salivary glands. Oncology (Williston Park) 1998;12:671–683.
32 Ellis GL, Auclair PL: Tumours of the Salivary Glands, ed 3. Washington, Armed Forces Institute of Pathology, 1996.
33 Gnepp DR, Brandwein MS, Henley JD: Salivary and lacrimal glands; in Gnepp DR (eds): Diagnostic Surgical Pathology of the Head and Neck. New York, Saunders, 1979, pp 408–429.
34 Renehan A, Gleave EN, Hancock BD, Smith P, McGurk M: Long-term follow-up of over 1,000 patients with salivary gland tumours treated in a single centre. Br J Surg 1996;83:1750–1754
35 McGurk M, Thomas BL, Renehan A: Extracapsular dissection, for clinically benign parotid lumps: reduced morbidity without oncological compromise. Br J Cancer 2003;89:1610–1613.
36 Thomas E, Hay EM, Hajeer A, Silman AJ: Sjögren's syndrome: a community-based study of prevalence and impact. Br J Rheumatol 1998;37:1069–1076.
37 Vitali C, Bombardieri S, Jonsson R, Moutsopoulos HM, Alexander EL, Carsons SE, Daniels TE, Fox PC, Fox RI, Kassan SS, Pillemer SR, Talal N, Weisman MH, European Study Group on Classification Criteria for Sjögren's Syndrome: Classification criteria for Sjögren's syndrome: a revised version of the European criteria proposed by the American-European Consensus Group. Ann Rheum Dis 2002;61:554–558.
38 Sjögren's International Clinical Collaboration Alliance (SICCA): http://sicca.ucsf.edu/
39 Daniels TE, Criswell LA, Shiboski C, Shiboski S, Lanfranchi H, Dong Y, Schiodt M, Umehara H, Sugai S, Challacombe S, Greenspan JS, Sjögren's International Collaborative Clinical Alliance Research Groups: An early view of the international Sjögren's syndrome registry. Arthritis Rheum 2009;61:711–714.
40 Shacham R, Droma EB, London D, Bar T, Nahlieli O: Long-term experience with endoscopic diagnosis and treatment of juvenile recurrent parotitis. J Oral Maxillofac Surg 2009;67:162–167.
41 Quenin S, Plouin-Gaudon I, Marchal F, Froehlich P, Disant F, Faure F: Juvenile recurrent parotitis: sialendoscopic approach. Arch Otolaryngol Head Neck Surg 2008;134:715–719.
42 Cascarini L, McGurk M: Epidemiology of salivary gland infections. Oral Maxillofac Surg Clin North Am 2009;21:353–357.
43 Brook I: The bacteriology of salivary gland infections. Oral Maxillofac Surg Clin North Am 2009; 21:269–274.
44 Barnes L, Eveson JW, Reichart P, Sidransky D (eds): World Health Organisation Classification of Tumours. Pathology and Genetics of Head and Neck Tumours. Lyon, IARC Press, 2005.
45 Larche MJ: A short review of the pathogenesis of Sjögren's syndrome. Autoimmun Rev 2006;5:132–135

46 Kassimos DG, Shirlaw PJ, Choy EH, Hockey K, Morgan PR: Chronic sialadenitis in patients with nodal osteoarthritis. Br J Rheumatol 1997;36:1312–1317.

47 McGurk M, Eyeson J, Thomas B, Harrison JD: Conservative treatment of oral ranula by excision with minimal excision of the sublingual gland: histological support for a traumatic etiology. J Oral Maxfac Surg 2008;66:2050–2057.

48 McGurk M: Management of the ranula. J Oral Maxfac Surg 2007;65:115–116.

49 Patel MR, Deal AM, Shockley WW: Oral and plunging ranulas: what is the most effective treatment? Laryngoscope 2009;119:1501–1509.

50 Sreebny LM, Schwartz SS: A reference guide to drugs and dry mouth, ed 2. Gerodontology 1997;14:33–47.

51 Nagler RM, Hershkovich O: Age-related changes in unstimulated salivary function and composition and its relations to medications and oral sensorial complaints. Aging Clin Exp Res 2005;17:358–366.

52 Scully C, Bagan JV, Eveson JW, Barnard N, Turner FM: Sialosis: 35 cases of persistent parotid swelling from two countries. Br J Oral Maxfac Surg 2008;46:468–472.

53 Bozzato A, Burger P, Zenk J, Uter W, Iro H: Salivary gland biometry in female patients with eating disorders. Eur Arch Otorhinolaryngol 2008;265:1095–1102.

54 Mandel L: An unusual pattern of dental damage with salivary gland aplasia. J Am Dent Assoc 2006;137:1498.

55 Kwon SY, Jung EJ, Kim SH, Kim TK: A case of major salivary gland agenesis. Acta Otolaryngol 2006;26:219–222.

56 Entesarian M, Dahlqvist J, Shashi V, Stanley CS, Falahat B, Reardon W, Dahl N: FGF10 missense mutations in aplasia of lacrimal and salivary glands. Eur J Hum Genet 2007;15:379–382.

Bethan L. Thomas
Guy's and St Thomas' NHS Foundation Trust
Department of Dental and Maxillofacial Radiology
Floor 23, Tower Wing, Guy's Hospital, London SE1 9RT (UK)
Tel. +44 207 188 1798, Fax +44 207 188 1873, E-Mail bethan.thomas@kcl.ac.uk

Author Index

Brown, J.E. 129

Carpenter, G.H. 107
Cotroneo, E. 107

DeVine, T. 48

Hoffman, M.P. 90

Larsen, M. 48
Lombaert, I.M.A. 90

McGurk, M. 129
Miletich, I. 1
Myat, M.M. 32

Patel, N. 78
Pirraglia, C. 32

Sequeira, S.J. 48

Thomas, B.L. 129
Tucker, A.S. VII, 21

Wells, K.L. 78

Subject Index

Adaptability, salivary glands 21–26, 109
AdCC, *see* Adenoid cystic carcinoma
Adenoid cystic carcinoma (AdCC),
extracellular matrix alterations 68, 69
Adherens junctions, lumen 86, 87
Apicobasal polarity, development in salivary
gland 87, 88
Apoptosis, lumen formation 80–83

Basement membrane (BM), *see also*
specific proteins
prospects for salivary gland studies 69, 70
proteins in submandibular gland
development 51–66
salivary gland function 49
Bioengineering, salivary gland
regeneration 121, 122
BM, *see* Basement membrane

Calculi, *see* Salivary gland calculi
Cancer, *see* Salivary gland tumors
Chondroitin sulfate proteoglycans, salivary
gland organogenesis role 59
cMyc, embryonic stem cell self-renewal
role 97, 98
Collagen, salivary gland organogenesis role
53, 54
Collagenase, salivary gland development
effects 53, 54

Dental glands, reptiles 27, 29
Drosophila salivary glands
anatomy 33
cell fate specification 33, 34
morphogenesis
axon guidance cues in cell
migration 42–44
cell migration 38–42
genes, table 35, 36
invagination 34, 36, 37
lumen elongation 37, 38
mutagenesis screens for gene
identification 44, 45
Dry mouth, *see* Xerostomia
Dystroglycan, salivary gland organogenesis
role 62

ECM, *see* Extracellular matrix
EGF, *see* Epidermal growth factor
Embryonic salivary gland stem/progenitor cells
basal progenitor cells 99, 100
mesenchymal growth factors and progenitor
cell fate 100–102
overview 95–97
transcription factors 97–99
Entactins, *see* Nidogens
Epidermal growth factor (EGF), salivary gland
organogenesis role 64
Epimorphin, salivary gland organogenesis
role 57
Extracellular matrix (ECM), *see also*
specific proteins
cell interactions in salivary gland
development 49, 50
neoplastic alterations in salivary glands
68, 69
organ culture studies of salivary gland
development 52, 53
prospects for salivary gland studies
69, 70
proteins in submandibular gland
development 51–66
salivary gland function 49
Sjögren's syndrome alterations 67

FGF, *see* Fibroblast growth factor
Fibroblast growth factor (FGF)
bud-related gene control 101
downregulation of embryonic stem cell maintenance genes 101
salivary gland
organogenesis role 63, 64
regeneration role 116, 118
Fibronectin, salivary gland organogenesis role 59, 60
Fine-needle aspiration (FNA), salivary gland tumors 139, 140
FNA, *see* Fine-needle aspiration

Heparan sulfate proteoglycans, salivary gland organogenesis role 58, 59

IGF, *see* Insulin-like growth factor
Insulin-like growth factor (IGF), salivary gland organogenesis role 64, 65
Integrins, salivary gland organogenesis role 60–62

Keratin, basal progenitor cell expression 99, 100
Klf4, embryonic stem cell self-renewal role 97
Knockout mice, salivary gland defect studies 9, 10

Laminin
distribution in salivary gland 55
knockout of laminin α_5 56
salivary gland organogenesis role 54–56
Ligation/deligation models, salivary gland duct 109–112
Lumen
apoptosis in space creation 80–83
elongation in *Drosophila* salivary gland development 37, 38
histology of formation 78–80
junctions in lining
adherens junctions 86, 87
apicobasal polarity role 87, 88
overview 83, 84
tight junctions 84–86
mouse hollow tube formation 16, 17
prospects for study 88

Major salivary glands, distribution 4, 5
MAPK, *see* Mitogen-activated protein kinase
Matrix metalloproteinases (MMPs)
salivary gland organogenesis role 65, 66
Sjögren's syndrome and extracellular matrix alterations 67
Minor salivary glands, distribution 4
Mitogen-activated protein kinase (MAPK), salivary gland regeneration role 117, 118
MMPs, *see* Matrix metalloproteinases
Mucous cell, features 3

Nest, salivary gland adaptation in bird nest formation 22, 23
Nidogens, salivary gland organogenesis role 56, 57
Notch, salivary gland regeneration role 118, 119

Obstructive disease, *see* Salivary gland calculi; Stricture
Oct3/4, embryonic stem cell self-renewal role 97
Organ culture, studies of salivary gland development 52, 53

p63, basal progenitor cell expression 100
PANS, *see* Parasympathetic autonomic nervous system
Parasympathetic autonomic nervous system (PANS), saliva production control 6, 7
Parenchyma, secretory endpieces 2
Parotid gland
development 8
features 4, 5
Perlecan, salivary gland organogenesis role 58, 59
Pilocarpine, xerostomia management 108
Proteoglycans, salivary gland organogenesis role 58, 59

Regeneration, salivary gland
bioengineering 121, 122
bioinformatics analysis of signaling pathways
mitogen-activated protein kinase 117, 118
Notch 118, 119
overview 118
Wnt 119
embryonic-like pathway 113–116
size maintenance of gland 112, 113
stem cell therapy 120, 121

Saliva
composition 2
functional overview 1, 2
production
control 6, 7
process 3
Salivary gland calculi
epidemiology 131
treatment 131, 133–135
Salivary gland stem cell (SGSC)
markers 94
pluripotency 94
Salivary gland tumors
adenoid cystic carcinoma and extracellular matrix alterations 68, 69
histology and classification 137, 138
incidence 136
management 137, 139, 140
site distribution 137
SANS, *see* Sympathetic autonomic nervous system
Serous cell, features 3
Sex differences, salivary glands 6
SGSC, *see* Salivary gland stem cell
Silk gland, salivary gland adaptation 21, 22
Sjögren's syndrome (SS)
classification 141, 142
epidemiology 141
extracellular matrix alterations 67
patient selection for clinical trials 142, 143
systemic effects 140
xerostomia pathogenesis 108
Sox
epithelial end bud expression 98, 99
Sox2 and embryonic stem cell self-renewal role 97
SS, *see* Sjögren's syndrome
Stem cell, *see* Embryonic salivary gland stem/progenitor cells; Salivary gland stem cell; Stem cell therapy; Submandibular gland
Stem cell therapy, salivary gland regeneration 120, 121
Striated duct, features 3
Stricture
epidemiology 131
treatment 135, 136
Sublingual gland
development 8, 12, 13
features 4, 5
Submandibular gland
development in mice
branch formation 14–16
hollow tube formation 16, 17
initial bud formation 11–13
overview 7, 8, 11, 91
epithelial stem/progenitor cells
adult cells 93–95
basal progenitor cells 99, 100
embryonic cells 95–100
mesenchymal growth factors and progenitor cell fate 100–102
overview 90–93
transcription factors 97–99
extracellular matrix and basement membrane proteins in development 48–66
features 4, 5
growth factors in development 96
Sympathetic autonomic nervous system (SANS), saliva production control 6, 7
Syndecan-1, salivary gland organogenesis role 62, 63

Tenascin C, salivary gland organogenesis role 57
Tetrodotoxin, salivary venom 23
TGF-β, *see* Transforming growth factor-β
Tight junctions, lumen 84–86
TIMPs, *see* Tissue inhibitors of matrix metalloproteinases
Tissue engineering, salivary glands 66, 67
Tissue inhibitors of matrix metalloproteinases (TIMPs), salivary gland organogenesis role 65
Transforming growth factor-β (TGF-β), basal cell regulation 101, 102

Venom, salivary gland adaptation in production
mammals 24, 25
overview 23, 24
reptiles
evolution of venom system 25, 26
teeth interaction with glands 27–29

Wnt, salivary gland regeneration role 119

Xerostomia
etiology 107, 108
ligation/deligation models 109–112
treatment 108